Advances in
Telecommunications Networks

For a complete listing of the *Artech House Telecommunications Library*,
turn to the back of this book

Advances in Telecommunications Networks

William S. Lee
Derrick C. Brown

Artech House
Boston • London

Library of Congress Cataloging-in-Publication Data
Lee, William S.
Advances in telecommunications networks / William S. Lee, Derrick C. Brown.
Includes bibliographical references and index.
ISBN 0-89006-606-X
1. Telecommunication systems.   I. Brown, Derrick C.   II. Title
TK5101.L397   1995   94-44499
621.382–dc20   CIP

A catalogue record for this book is available from the British Library
Lee, William S.
Advances in Telecommunications Networks
I. Title   II. Brown, Derrick C.
621.382

ISBN 0-89006-606-X

© **1995 ARTECH HOUSE, INC.**
685 Canton Street
Norwood, MA 02062

International Standard Book Number: 0-89006-606-X
Library of Congress Catalog Card Number: 94-44499

10   9   8   7   6   5   4   3   2   1

Contents

Preface

The purpose of this book is to provide insight into the many advances in long-distance telecommunications that have occurred in recent years. The technology used in long-distance networks has changed from analog to digital, and the predominant transmission medium has changed from microwave radio to fiber-optic cable.

The authors have been involved in many of the developments in network elements and transmission systems as they have evolved, and they have endeavored in this book to compile information that will be useful to those desiring a broad understanding of network technology and services. It is recognized that the technology is constantly evolving, and that many new standards are being published every year. For these reasons, no attempt has been made to give fine details of topics that are likely to change in the near future. The authors' approach has been to present the principles involved in the operation of equipment and systems and to give some outlook on the directions in which networks and services will evolve in the coming years.

This book is intended for the those engaged in overseeing, planning, designing, or operating a digital network. The field of communications transmission is composed of a central core of specialized knowledge and design practices, derived from a broad scientific and engineering base built up over a long period of time. The first few chapters of the book review a subset of that knowledge needed to follow the book.

The authors have aimed at presenting the materials from a practical standpoint, avoiding mathematical equations. The presentation is that of an overview of the modern communications network, which will be useful to the communications manager, while providing enough technical information for those seeking more detailed coverage of various aspects of the subject. It is expected that those wishing to seek even more details will consult the publications of the various standards bodies referenced in the bibliography.

Acknowledgment is made to our colleagues at the Sprint Corporation for their support and encouragement, and in particular to Terry Kero, director of the Advanced Technology Laboratories in Burlingame, California.

Introduction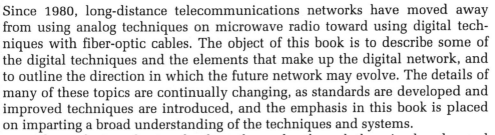

Since 1980, long-distance telecommunications networks have moved away from using analog techniques on microwave radio toward using digital techniques with fiber-optic cables. The object of this book is to describe some of the digital techniques and the elements that make up the digital network, and to outline the direction in which the future network may evolve. The details of many of these topics are continually changing, as standards are developed and improved techniques are introduced, and the emphasis in this book is placed on imparting a broad understanding of the techniques and systems.

The evolution of networks, from the analog through the mixed analog and digital phase to the current all-digital configurations, forms the subject of Chapter 2. This is followed by a review of digital multiplexing and transmission in Chapter 3 and descriptions of the digital signals in Chapters 4, 5, and 6. Methods of synchronization of signals and networks are covered in Chapter 7.

Digital cross-connect systems (DCS) are an important part of the technology of the networks, and the DCS 1/0, DCS 3/1, and DCS 3/3 are described in Chapters 8, 9, and 10. The network operations aspects are covered in Chapter 11, and the use of the DCS 3/3 system in the survivability of traffic in the event of a failure in the network is described in Chapter 12.

Synchronous networks are being introduced with the synchronous optical network (SONET) technology described in Chapter 13, and the use of rings made up of SONET systems to achieve self-healing networks is covered in Chapter 14.

Finally, the direction in which networks may evolve to handle future broadband services is discussed in Chapter 15.

Network Architecture Evolution

2.1 EVOLUTION FROM ANALOG TO DIGITAL NETWORKS

2.1.1 Analog Networks

Long-distance telecommunications networks used analog techniques for several decades until around 1980. Traffic was principally speech, with some low-speed data. The end-to-end composition of a long-distance circuit included signaling, multiplex, cable or radio line transmission equipment, and, unless it was used for private leased line service, a toll switch. A typical circuit block schematic is illustrated in Figure 2.1. The traffic that passes between the network and the local telephone office, which is called access traffic, was carried on balanced pair cables using voice frequency (VF) signals.

Steady advances in component technology allowed vast improvements in system design and the equipment packaging densities required, and power consumption decreased substantially when semiconductors replaced tubes. However, the block schematic of the analog circuit remained virtually unchanged for over fifty years. The basic disadvantages of analog systems, including the accumulation of noise with circuit length, as well as the special conditioning needed for any traffic other than speech and voiceband data remained major problems.

2.1.2 Mixed Analog and Digital Networks

The first application of digital techniques was to reduce the number of balanced cable pairs needed when interconnecting local-area telephone offices, using the 24 channel T1 system with PCM channel banks at each end. This was a short-distance application, and all channels, analog and digital, still had VF interfaces.

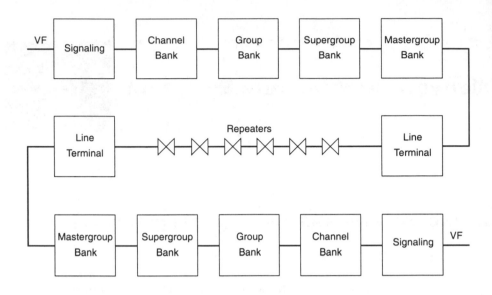

Figure 2.1 Analog circuit block schematic.

The introduction of digital ports at the DS1 rate in place of VF ports on toll-switching equipment affected the economics of interconnecting the switches and the multiplex equipment. PCM channel banks were used back-to-back with analog channel banks, interconnected at a VF patching frame, as shown in Figure 2.2. The cost of the channel-bank equipment, as well as the frequency response and group-delay distortion introduced by this arrangement, were substantial disadvantages. Some vendors developed combined PCM and analog channel banks, which removed VF circuitry duplicated in the back-to-back arrangement and reduced the performance impairments.

The access traffic between the local telephone offices and the long-distance networks had been carried at the VF level on balanced-pair cables, but more cost-effective digital interconnections at the DS1 level began being introduced in 1983. The interconnection of these access facilities and the analog channel banks also required the back-to-back channel bank arrangements.

Transmultiplexers were developed as another solution to the analog/digital interface problem. The group transmultiplexer transformed one 24-channel DS1 signal into two 12-channel analog groups, and vice versa. A supergroup transmultiplexer transformed five DS1 signals into two 60-channel analog supergroups. The first of these eliminated the back-to-back channel banks, and the second also removed the group bank. These arrangements allowed the digital DS1 interfaces at the toll switch and access traffic points to be connected to

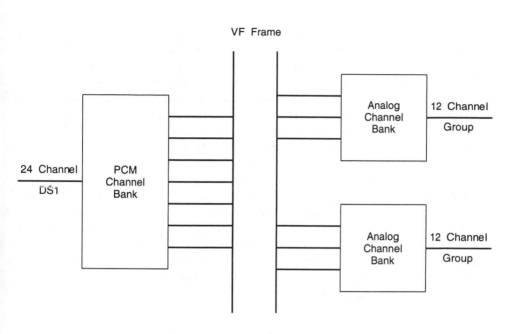

Figure 2.2 Back-to-back channel banks.

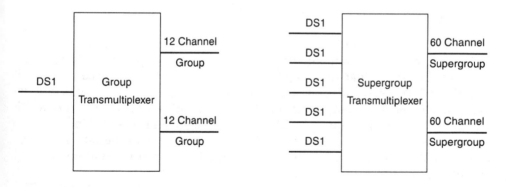

Figure 2.3 Group and supergroup transmultiplexers.

the higher order analog multiplex equipment, as shown in Figure 2.3. The transmultiplexers converted analog signals into digital signals, but they handled only blocks of voice frequency channels, not wideband digital signals.

2.1.3 Digital Networks

Soon after 1980, the rapid advances in digital technology made an all-digital network economically feasible. Digital technology offered superior service quality for voice traffic and made possible new high-speed data services. Digital transmission systems using fiber-optic cables, with a capacity of thousands of channels, were introduced in the early 1980s. This development allowed the mixed analog/digital arrangements to be superseded by fully digital systems. Existing analog transmission systems were taken out of service and replaced by digital networks. The PCM channel bank, which includes the signaling function, and the M1-3 multiplexer replaced the analog multiplexing equipment. The trans-multiplexers, having played their part in the transition, were phased out as the digital transmission systems were brought into service.

The introduction of remotely controlled digital cross-connect systems (DCS) allowed digital networks to be monitored and controlled from a centralized operations center. Many tasks which used to be performed manually by personnel at the network sites can now be performed remotely. Reconfiguration that allows for changes in traffic demand or that overcomes a failure in some part of the network may be carried out very rapidly. Synchronous optical network (SONET) transmission systems, which allow remote performance monitoring of the network and self-healing techniques for network survivability, are also being introduced.

2.2 ASYNCHRONOUS DIGITAL NETWORKS

2.2.1 General

The digital networks initially deployed for long-distance service have the clocks at each network switching site synchronized to a primary timing reference, but the traffic that they carry is asynchronous. This is because the bit rates of the DS1, DS2, and DS3 digital signals are allowed to vary within a defined range on each side of the nominal values. In addition, because the multiplexing process that generates the DS2 and DS3 signals inserts stuffing bits into the pulse stream, it is not possible to locate the positions of the bits belonging to an individual DS0 or an individual DS1 in a DS3 signal without completely demultiplexing the DS3 stream.

No provision is made in these systems for remote equipment provisioning, but a limited amount of performance monitoring may be provided on some DS1 and DS3 facilities. The coding of the digital traffic onto the optical path is proprietary to each vendor, and systems with the same nominal bit rate cannot be interconnected at the optical level on a fiber cable route.

2.2.2 Configuration with VF Interfaces

When the access-traffic interface was at the VF level, the basic block schematic of a digital circuit was simple, as illustrated in Figure 2.4. Every channel was able to carry any of the available types of service, and the channel bank was

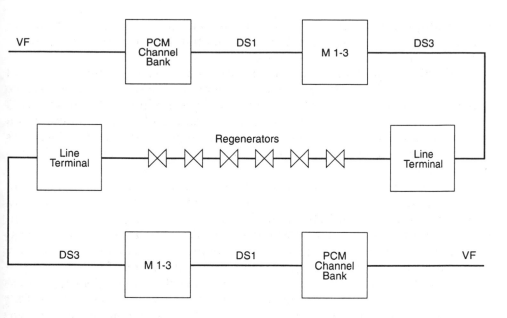

Figure 2.4 Basic digital circuit block schematic.

equipped with the plug-in unit containing the correct signaling option for the traffic to be carried on that channel. Each VF access circuit was multiplexed into its assigned allocation in a DS3 facility traveling towards its destination, and equipment interconnections at the VF, DS1, and DS3 levels were performed manually using VF, DSX-1, and DSX-3 frames.

2.2.3 Configuration with Digital Access Interfaces

The introduction of DS1 access interfaces may make the multiplexing arrangements for individual circuits more complicated, because the overall configuration of a circuit depends on the type of service being carried.

Switched traffic is handled at the DS1 level, rather than at the individual circuit level. It is routed to a DS1 line port by the toll switch and then is connected to an M1-3 multiplexer, which in turn connects to the appropriate line system. Private line circuits are established on a long-term basis between two sites on the network and therefore do not pass through toll switches. An access DS1 facility is likely to include private-line circuits having different destinations and requiring a variety of routings. The circuits may be reassigned using PCM channel banks on a back-to-back basis and cross-connecting the circuits at the VF frame. By this means, one DS1 access facility may have its private line circuits rerouted to a number of network DS1 facilities, which are connected via

M1-3 multiplexers to line systems. The rearrangement of circuit allocations among multiplex facilities, in order to reroute traffic or to reduce the quantity of facilities needed, is called grooming.

A further stage in the reduction of access facility cost was the provision of a DS3 access interface between the local telephone office and the long-distance network. The multiplexing of the switched and private line traffic onto the DS3 access facilities is performed in the local area and tends to use the minimum quantity of DS3 facilities necessary to carry the total access circuits. The access facilities are then groomed in the long-distance network to meet network facility and traffic plans.

The configuration of a site that includes a toll switch but no cross-connect systems is shown in Figure 2.5. The access traffic enters on DS3 interfaces, which are terminated on M1-3 multiplexers. The DS1 facilities that carry private-line traffic pass through PCM channel banks, which are connected back-to-back at the VF level to allow traffic grooming. The DS1 blocks that carry switched traffic connect to the drop side of the toll switch. The groomed private-line traffic and the DS1 signals from the line side of the toll switch are assigned to M1-3 multiplexers, which connect to the line systems at the DS3 level. For the sake of simplicity the DSX-1 and DSX-3 frames are not shown, but every cross connection on these frames is made using wire jumpers, and no remote control of these interconnections is possible.

2.2.4 Introduction of Digital Cross-Connect Systems

The first of the cross-connect systems to be deployed in networks was the DCS 1/0, and it replaced the back-to-back PCM channel banks in the grooming of private-line traffic. In addition to providing remotely controlled cross connection of DS0 signals, it allows the testing of the circuits from a central network location.

The DCS 3/3 system was the next to be introduced. Its initial application was network survivability, but it is also used to provide remote DS3 facility provisioning, which allows deletion of DSX-3 frames.

The DCS 3/1 is the last of the DCSs to be developed and introduced into long-distance networks. With its ability to provide grooming of DS1 streams, it can replace the M1-3 multiplexer and the DSX-1 frame, and when the "DS3 Intact" feature is used the DSX-3 frame can also be deleted.

The configuration of a toll switch site when all three types of cross-connect systems are deployed is shown in Figure 2.6. The DCS 3/1 has both DS1 and DS3 ports, and it allows the private line and switched traffic to be segregated at the DS1 level. Since remotely controlled cross-connection capability is provided at the DS0, DS1, and DS3 levels, the addition, deletion, or rearrangement of network facilities can be handled without requiring personnel at the site.

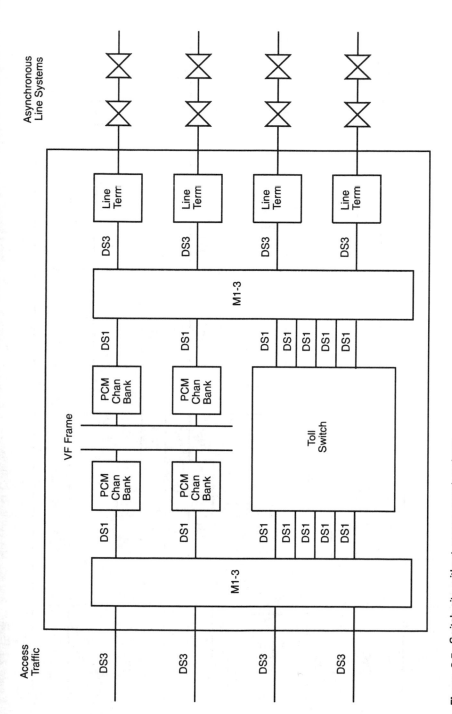

Figure 2.5 Switch site without cross-connect systems.

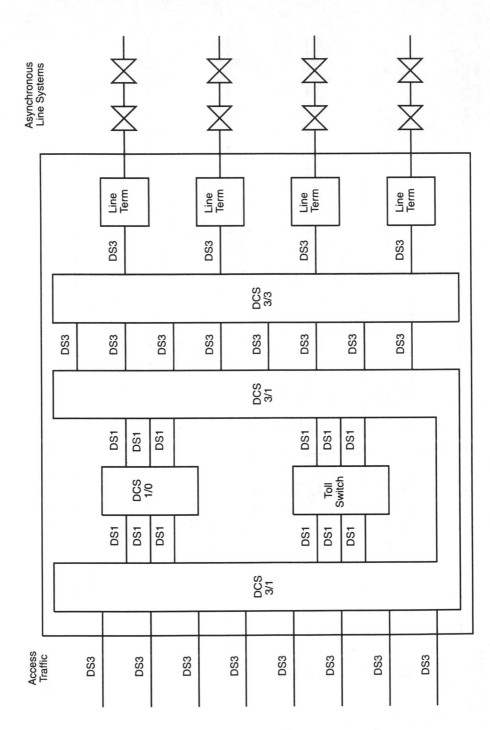

Figure 2.6 Switch site with cross-connect systems.

2.2.5 Intermediate Access Sites

At access sites located along a line system between major sites, back-to-back line terminals are used to provide DS3 interfaces, as Figure 2.7 shows. The local access traffic interface is at the DS3 level and is connected to the line terminal via the reverse direction protection switch (RDPS), as detailed in Chapter 12. No facility grooming is performed at the access site; this is handled by the DCS 3/1 and DCS 1/0 systems at the major site on which the traffic is homed.

2.3. SONET NETWORKS

2.3.1 General

The adoption of the SONET technique allows completely synchronous networks to be established. As explained in Chapter 13, the electrical signal at the lowest level in the SONET hierarchy is the STS-1, which has a rate of 51.84 Mbps. The rates of the higher level STS-N signals are multiples of the STS-1 rate, where N is a whole number. The light signals on the fiber-optic cables are the optical equivalents of the STS-N, and are denoted by OC-N. Standardized line rates and optical signal formats allow the end-to-end interconnection of equipment supplied by different vendors.

SONET systems carry asynchronous facilities, such as DS1 and DS3 signals, by mapping them into synchronous payloads. The DS3 is mapped into the payload capacity of one STS-1, but lower speed traffic, such as the DS1, is mapped first into a virtual tributary (VT) format. Individual virtual tributaries are collected into VT groups, and the groups are then mapped into an STS-1 payload. Optical interconnection of SONET network elements within a site reduces the cost and complexity involved in implementing similar interconnections at the DS3 level. These optical interconnections are also synchronous, allowing high speed traffic to be carried intact through a site from one system to another. Optical access points for these high-speed services have to be provided at access sites, in addition to the metallic DS1 and DS3 interfaces.

SONET networks include add/drop multiplexers (ADM), which allow DS1, DS3, and OC-N blocks to be added and dropped at intermediate sites along a line system. Remote control of SONET network elements, including cross-connect systems, supplies the means to provide equipment and facilities and retrieve performance-monitoring data.

2.3.2 Site Configuration

Figure 2.8 shows the configuration of a SONET network switch site and may be compared with the equivalent site in the asynchronous network shown in Figure 2.6. The cross-connect systems perform the same grooming functions, but

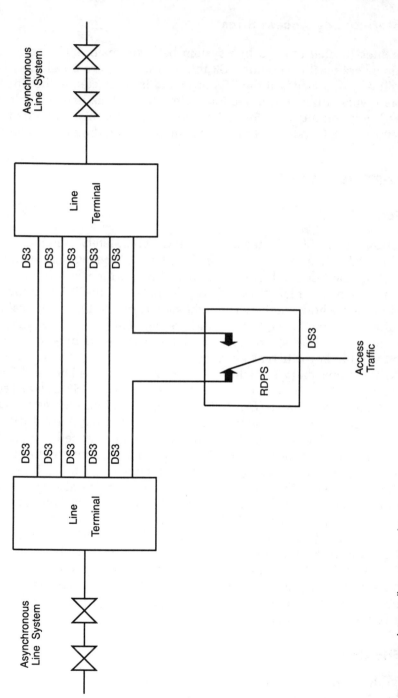

Figure 2.7 Intermediate access site.

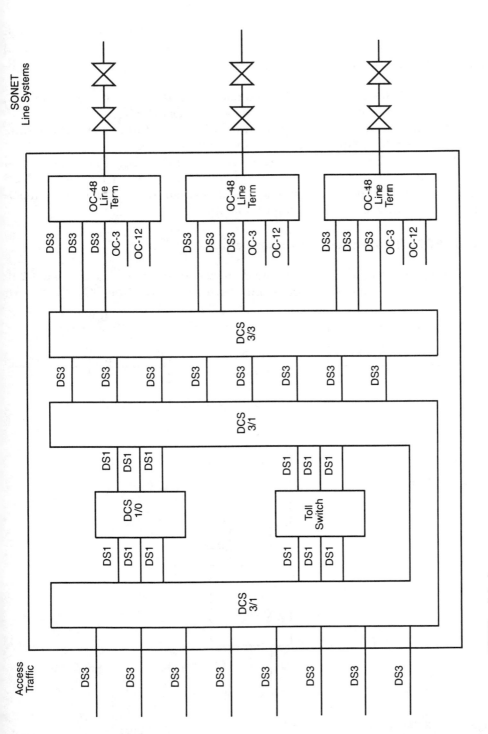

Figure 2.8 SONET switch site.

the OC-48 line terminals have OC-3 and OC-12 optical drop interfaces in addition to the DS3 ports. These optical ports may be used to connect high-speed traffic from one line system route to another, or as access ports for local high-speed services.

The DCS 3/1 and DCS 3/3 cross-connect systems have matrix speeds, which allow the cross connection of signals at SONET rates. It has been proposed that these two systems may be fitted with optical interfaces, thus reducing the number of metallic DS3 interconnections at a site.

Where SONET rings are implemented for survivability, the loop reversal survivability technique using the DCS 3/3 becomes obsolete. However, the DCS 3/3 equipment would still function as a remotely controlled frame for connecting, disconnecting, and rearranging DS3 facilities.

2.3.3 SONET Intermediate Access Sites

The configuration of a SONET intermediate access site is shown in Figure 2.9. An ADM is used in an OC-48 route and provides access at the OC-12 level. An

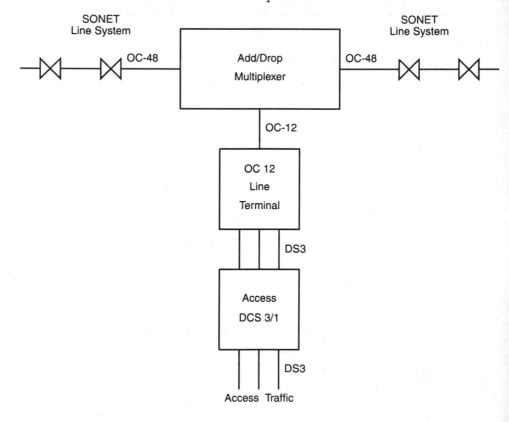

Figure 2.9 SONET intermediate access site.

OC-12 line terminal is used to provide a DS3 level interface point, and a small capacity access DCS 3/1 system is used to groom the DS3 access traffic.

This arrangement provides advantages over its counterpart in the asynchronous network. The access traffic may be groomed into facilities assigned to their network destinations, and more than one OC-12 may be dropped at the site when the volume of traffic justifies it. The traffic may be redirected using the ADM in the event of a cable break on either side of the access site, and all of the equipment is remotely controllable for reconfiguration purposes. Any problems with through or access traffic may be analyzed using the performance monitoring features, allowing the fault condition to be identified.

Digital Multiplexing and Transmission

3

3.1 INTRODUCTION

Before new network elements and systems are discussed in detail, this chapter presents a brief review of digital multiplexing and transmission. This discussion does not cover SONET techniques, as Chapter 13 details this subject.

3.2 DIGITAL SIGNALS

Digital transmission is characterized by the use of a train of pulses, which may have positive and negative values, and spaces with zero amplitude between the pulses. All of the pulses are the same height, which contrasts with analog transmission, where the instantaneous amplitude of the signal corresponds directly to the magnitude of the information. When digital systems first appeared in networks, the principal traffic was speech. Low-speed data were converted into voice frequency tones to be carried over the systems. Later developments allowed data traffic to be carried digitally at various bit rates. The technique used to convert speech signals into digital pulse streams is called pulse code modulation (PCM).

3.3 PULSE CODE MODULATION

Early forms of communication employed patterns of pulses of different lengths to represent letters of the alphabet. One example of this is Morse code, used over wire and radio telegraph systems. That technique involves the encoding of a written message letter by letter and does not transform speech directly into a digital format. This is accomplished by PCM [1].

PCM was invented in 1938, but only the advent of the semiconductor allowed practical systems to be built. PCM encodes the instantaneous value of an analog signal, rather than an individual letter of a word. Samples of the signal's amplitude are taken, as shown in Figure 3.1. The sampling frequency must be at least twice that of the highest frequency component of the signal being sampled to allow all of the information to be recovered correctly at the receive end of the system. Speech signals are filtered to limit the top frequency to 3,400 Hz, and the PCM systems use a sampling frequency of 8 kHz—more than twice the highest frequency of the signals—resulting in a sampling interval of 125 microseconds.

Prior to encoding, each sample is quantized, which means that its amplitude is measured against a scale to determine its value. If a linear quantizing scale were used, a very large number of values would be encoded because speech has a wide range of amplitudes. However, most speech samples have a small amplitude, and a non-linear scale is used to assign most of the steps to these low-amplitude samples. Each sample is compared with the quantizing curve and assigned the value of the closest step on the scale.

The first North American PCM systems encoded the speech samples into seven bits and used an eighth bit for transmitting signaling information. Subsequent improvements resulted in 8-bit encoding for five out of six successive samples, with 7-bit encoded speech and a signaling pulse being used in the sixth case. Seven-bit encoding allows 127 quantizing steps to be used, whereas eight bits can accommodate 255 steps, which allows more accurate quantizing of the speech samples. When the 8-bit binary code is used, the value of each bit position assigned is as illustrated in Figure 3.2, which also shows the bit streams for various amplitude values. The block of eight bits is called a byte.

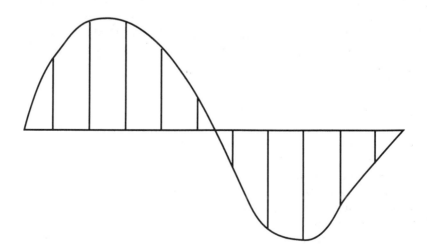

Figure 3.1 Amplitude sampling.

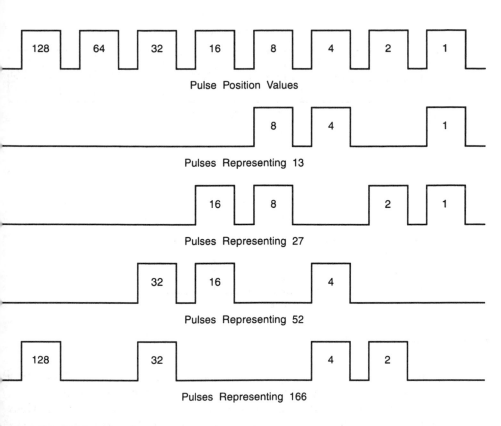

Figure 3.2 PCM pulse trains.

3.4 DIGITAL MULTIPLEXING

The North American PCM channel bank provides digital multiplexing [2] capacity for 24 channels, any of which may have voice frequency or digital data input interfaces. The 8-bit bytes from each channel, which constitute the DS0 signal, are transmitted in turn for a total of 192 bits, as illustrated in Figure 3.3. In order to mark the block of 24 channels, an additional bit, called the framing bit, is added, so each frame consists of 193 bits. Each 24-channel frame occupies the sampling interval of 125 microseconds, corresponding to a bit rate of 1.544 megabits per second (Mbps), and is known as the DS1 signal. The exact bit rate depends on the accuracy of the frequency of the 8-kHz signal used in the sampling process, and the 1.544-Mbps rate is allowed to vary plus or minus 32 parts per million.

The second-stage multiplexer, called the M1-2, combines four DS1 signals into the DS2 format at 6.312 Mbps, interleaving the four tributary DS1 signals in turn. Since each of the DS1 bit streams is allowed to vary from the nominal

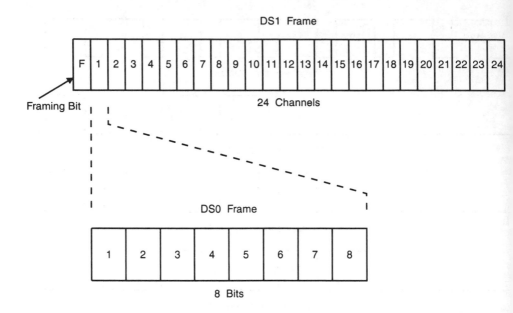

Figure 3.3 DS1 frame structure.

rate, dummy stuffing pulses are added as needed to bring the combined speed up to 6.312 Mbps, with a tolerance of plus or minus 33 parts per million.

The multiplexing together of seven DS2 streams by sequential interleaving in the M2-3 multiplexer, which also uses stuffing pulses, produces the DS3 signal at 44.736 Mbps, with a tolerance of plus or minus 20 parts per million. Since the DS2 signal is not generally employed in transmission systems, a combined M1-2 and M2-3 instrument called the M1-3 has become the standard multiplexer.

A fourth step, which combines six DS3 streams to produce the 274.176-Mbps DS4 rate, was used on the first North American coaxial cable system, but this step has not been used in the more recent transmission applications.

Various other higher speed signals have been developed for use on radio and fiber-optic systems, e.g., 135, 417, and 565 Mbps and 1.12 and 1.7 Gbps. These employ proprietary modulation schemes developed by system manufacturers, and the pulse stream formats are not standardized.

3.5 DIGITAL HIERARCHIES

Three hierarchies of PCM systems have evolved as shown in Table 3.1. The hierarchy used in Japan is identical with the North American version only for the DS1 and DS2 levels and has different rates for the higher stages. The North American and CEPT (Conference of European Posts and Telegraphs) versions have only the single channel bit rate in common. The CEPT and the North American systems use different quantizing scales, called A-Law for the CEPT and μ-Law for the North American system, and thus no direct interconnection of digital signals is possible between the two hierarchies.

Table 3.1
Hierarchies of PCM Systems

Hierarchy Level	Parameter	North America	CEPT	Japan
0	Designation	DS0		DS0
	Bit Rate Kbps	64	64	64
	VF Channels	1	1	1
1	Designation	DS1	E1	DS1
	Bit rate Mbps	1.544	2.048	1.544
	VF Channels	24	30	24
	Companding Law	μ	A	μ
1C	Designation	DS1C		
	Bit Rate Mbps	3.152		
	VF Channels	48		
	Companding Law	μ		
2	Designation	DS2	E2	DS2
	Bit rate Mbps	6.312	8.448	6.312
	VF Channels	96	120	96
3	Designation	DS3	E3	
	Bit rate Mbps	44.736	34.368	32.064
	VF Channels	672	480	480
4	Designation	DS4	E4	
	Bit rate Mbps	274.176	139.264	97.728
	VF Channels	4032	1920	1440
5	Designation	Not	E5	
	Bit rate Mbps	defined	565.148	397.200
	VF Channels		7680	5760

A standard has been adopted to allow transmission of both North American- and CEPT-based traffic on the North Atlantic digital cable routes. The 139.264-Mbps rate, which was evolved to carry four 34.368-Mbps CEPT signals, can also handle three 44.736-Mbps North American bit streams. It is then referred to as the DS4E level.

The DCSs described in this book are all based on the North American hierarchy, and only passing reference to the CEPT standards is included in the remaining chapters.

3.6 SYNCHRONIZATION

The DS1, DS2, and DS3 bit streams have a tolerance on their bit rates, as mentioned in the section on multiplexing. In order to identify and recover each individual channel correctly at the receive end of a system, the timing clock used to demultiplex the received signals must be derived from the incoming bit stream. This is the bit-synchronization process. The methods used to synchronize the transmitting clocks are discussed in Chapter 7.

Stuffing pulses are used at each multiplexing level to allow for the bit-rate tolerances of individual digital traffic streams. In the demultiplexing process these stuffing pulses have to be identified and removed. The number of these stuffing pulses in the higher rate bit streams vary, so the exact locations of each of the 28 DS1s in a DS3 signal is unknown. Thus, although the clocks at the terminals at the ends of a system are synchronized with one another, there is no means of extracting a DS1 signal at any intermediate site without demultiplexing the complete higher speed bit stream.

3.7 REGENERATION

Signals in a transmission system are attenuated as they pass along the cable or radio path. In analog applications, repeaters are provided at points along the path to amplify the signals. The repeaters also amplify the noise that accompanies the signals and add idle and intermodulation noise of their own, so that noise increases along the length of the system.

The pulses in a digital transmission system become distorted as they pass along the path. Instead of amplifiers, a digital system uses regenerators, which accept distorted pulses and provide new undistorted pulses at the output. There is no accumulation of noise on a digital system.

3.8 TRANSMISSION SYSTEMS

Four types of transmission media are used for carrying digital traffic—metallic pair cable systems, coaxial cable systems, microwave radio systems, and fiber optic cable systems—and each is described below.

3.8.1 Metallic Pair Cable Systems

The initial digital system employed the first-level rate of 1.544 Mbps and used balanced pair cables. This system, known as the T1, often used existing cable plant that had previously been used to carry voice frequency traffic between telephone offices. By providing 24 channels in place of one, it offered great savings by avoiding investment in new cable plant. Later, the 48-channel T1C system, operating at the 3.152-Mbps DS1C rate, enabled even more economical utilization of the cables. However, these systems were limited in length by the accumulation of interference due to crosstalk between the cable pairs. They were usually applied on routes up to 25 miles in length.

The DS2 rate system, called the T2, was used very little because the crosstalk and other limitations of normal balanced pair cables caused too many problems at the 6.312-Mbps rate. Special cables with greatly improved performance characteristics were developed to support the T2 system, but the increased cable cost made most applications uneconomic.

3.8.2 Coaxial Cable Systems

Coaxial cables allow much higher bit rates to be carried than could be accommodated on balanced pairs. The first of these coaxial cable systems to be introduced was a 274.176-Mbps version put into service in Canada. Coaxial cables were already in extensive use for analog transmission, with bandwidths up to 60 MHz, although the majority of the systems deployed in Europe had top frequencies of 4 or 12 MHz. The analog systems have been replaced extensively by the European standard 139.264-Mbps digital system.

3.8.3 Microwave Radio Systems

Unlike Europe, where radio system use is severely limited by path congestion and interference due to the close proximity of population centers, the long-distance service in North America has until recently used microwave radio as the principal transmission medium. Conversion of these radio routes from analog to digital traffic started in the late 1970s. Various ingenious and complex techniques have been developed to code the digital signals in an effort to make them fit within the radio channel bandwidths allowed by the Federal Communications Commission (FCC). The coding improvements have been directed at increasing the digital bit rate carried within a fixed megahertz limit, but the more bits carried per hertz, the more susceptible the signals are to noise and other interference. Major microwave-radio routes, which are capable of carrying 2,700 analog channels, can support only 3 DS3 streams, which comprise 2,016 digital channels.

3.8.4   Fiber-Optic Cable Systems

The announcement by Corning in 1970 of the development of an optical fiber with a loss of only 20 dB/km was the beginning of the introduction of a totally new transmission medium. Unlike existing copper pair and coaxial cables, fiber-optic cables use low-cost, abundantly available materials. In addition, fiber-optic cables have very wide bandwidth, no crosstalk problems, and are light and very small. The fiber attenuation has now been reduced to less than 1 dB/km. The continuing development of optical components, including light-emitting diodes (LED), lasers, and optical detector elements, allows very high bit rates to be achieved. Other necessary techniques have evolved, including the fabrication of multi-fiber cables and fiber-jointing techniques that can be used under field conditions.

The North American digital hierarchy has been standardized only up to the DS3 level. Various optical-line rates have been used by manufacturers for their proprietary fiber-cable systems, but none of these offers any compatibility above the DS3 level. As a result, the same vendor's equipment must be used at both ends of these systems, and access to the traffic at any point along a route requires back-to-back terminals, which demodulate all of the traffic down to the DS1 or DS3 levels. As a means of overcoming these limitations, and to provide synchronous transmission, SONET has been developed. SONET is discussed in Chapter 13.

3.9   CROSS-CONNECT FRAMES

In order to allow flexibility of interconnection of digital equipment, cross-connect frames are used at the voice frequency (VF), DS1, and DS3 levels. Figure 3.4 shows a typical arrangement, with PCM channel banks, M1-3 equipment, and transmission systems interconnected with frames. The frame at the DS1 level is called the DSX-1, and the frame at the DS3 level is the DSX-3. For both frames there are standard limits on the pulse rate, pulse shape, and power level of the digital streams and on the length of cable allowed between the frame and the equipment connected to it [3]. Access points are provided to enable monitoring of each cross connection without affecting traffic.

The VF frame uses 600-ohm balanced pair; the DSX-1 frame uses 100-ohm balanced pair; and the DSX-3 has 75-ohm unbalanced coaxial pair connections. In each case cross connections are made manually with jumper cables. At sites where frequent and extensive rearrangements are necessary, changing the jumper positions is time-consuming and can result in connection errors and loss of service before the mistakes are identified and corrected. In addition, network facility rearrangements may involve coordinated action at a number of sites, further complicating the activities. With the excellent reliability of modern electronic equipment, many sites are normally not staffed, so

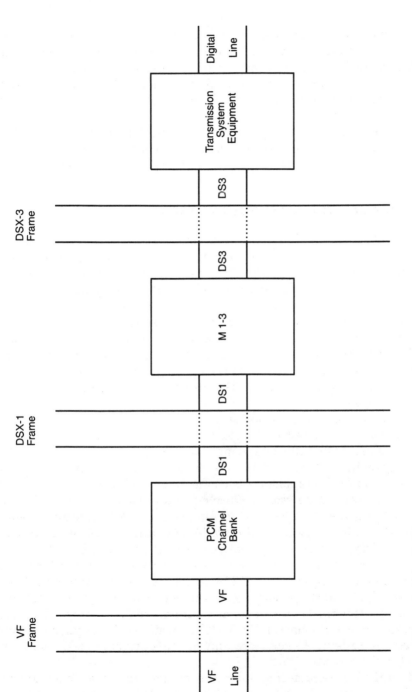

Figure 3.4 Cross-connection frames.

cross-connection frame activities involve sending personnel to each location. Failures of transmission links require extensive temporary rearrangements to restore traffic, and manual jumpers on the frames may take hours to implement, with considerable loss of traffic. After the network is repaired, all of the jumpers must be restored to their normal configurations, providing more possibilities of connection errors.

All of these factors mean that there is a considerable delay in the implementation of any request for adding, deleting, or changing service on a network, when cross connections at the VF, DSX-1, and DSX-3 frames have to be made.

3.10 DIGITAL CROSS-CONNECT SYSTEMS

The DCSs perform many functions [4,5]. One of these is to enable cross connections at the DS0, DS1, and DS3 levels to be added, deleted, and changed by remote control from a distant site. The first DCS is the DCS 1/0, which provides cross connections at the DS0 rate and has DS1-rate external connections. This system is described in Chapter 8. Any DS0 in an incoming DS1 stream may be assigned into any vacant DS0 slot in an outgoing DS1 signal. In addition, a complete incoming DS1 may be connected to any vacant outgoing DS1 port. This is called *DS1 intact* cross connection, and it is this feature that allows the DCS 1/0 to act as a remotely controlled replacement for the DSX-1 manual frame. However, because the DS1 intact connection uses 24 DS0 matrix cross connections in each direction of transmission, this is an inefficient use of a system primarily intended for DS0 service.

The DCS 3/1, described in Chapter 9, cross connects DS1 streams and has DS1- and DS3-rate external connections. Any DS1 in an incoming DS3 stream may be assigned into any vacant DS1 slot in an outgoing DS3 signal, and provision is made also for DS3-intact cross connection. In addition, any signal at an incoming DS1 port can be cross connected to any vacant outgoing DS1 port. The DCS 3/1 can therefore be used to replace the M1-3 multiplexer and the DSX-1 and DSX-3 frames, and all of these functions are remotely controllable.

The third type of cross connect is the DCS 3/3, which is described in Chapter 10. This is less complex than the two previous types and cross connects DS3 signals. It employs only DS3-rate external connections and therefore provides only DS3 intact operation. At a site where there is no DS1 traffic, the DCS 3/3 can replace the DSX-3 frame much more efficiently than a DCS 3/1. A digital cross-connect system 1/1 (DCS 1/1) is available for applications where DS1 signals are to be cross connected intact. This system is less widely used than the other DCS types.

Each of the cross-connect systems may be used to replace manual frames, and the DCS 3/1 may also take the place of the M1-3. The ease with which digital traffic may be added, deleted, and rerouted by remote control is one of the great benefits resulting from the use of these systems.

3.11 TRAFFIC ASSIGNMENT AND GROOMING

Economic operation of a network requires that transmission systems and other network elements are utilized efficiently. A measure of efficiency is the extent to which the capacity of DS1 and DS3 facilities is filled with traffic. Traffic is assigned to the most direct routes wherever possible, and grooming is carried out to improve the fill factor of the facilities. For example, if a DS1 facility with a capacity of 24 channels carries only 8 channels, the fill factor is 1/3. If by traffic grooming the facility can be assigned 16 channels, the fill factor is improved to 2/3.

In Figure 3.5, five underutilized DS1 facilities are shown at the left-hand side of the drawing. Five PCM channel banks are used to break these facilities down to the VF channel level. By cross connection on the VF frame, the VF channels are reassigned onto three PCM channel banks so that the five original DS1 facilities have been groomed onto three DS1 facilities. The same grooming functions may be performed using the DCS 1/0 system, as illustrated in Figure 3.6. Comparing Figures 3.5 and 3.6, the DCS 1/0 replaces all the PCM channel banks and the VF frame and offers remote control of the cross-connection process.

Grooming at the DS3 level with M1-3 multiplexers and the DSX-1 frame is shown in Figure 3.7. Five partly filled incoming DS3 facilities are groomed onto three outgoing DS3 streams. The use of the DCS 3/1 to perform the same grooming is illustrated in Figure 3.8, and the DCS 3/1 replaces all of the M1-3 multiplexers and the DSX-1 frame, in addition to providing remote control of the grooming operation.

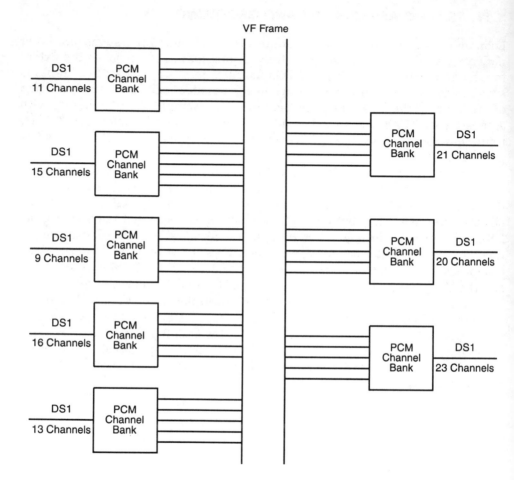

Figure 3.5 DS1 grooming with PCM channel banks.

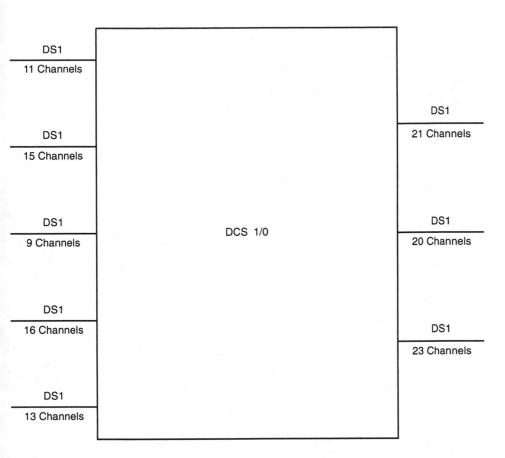

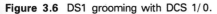
Figure 3.6 DS1 grooming with DCS 1/0.

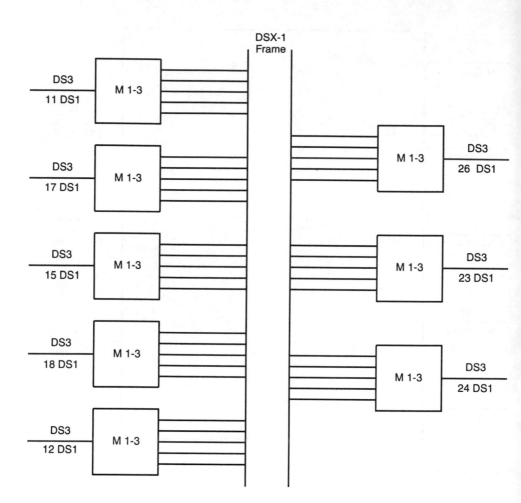

Figure 3.7 DS3 grooming with M1-3.

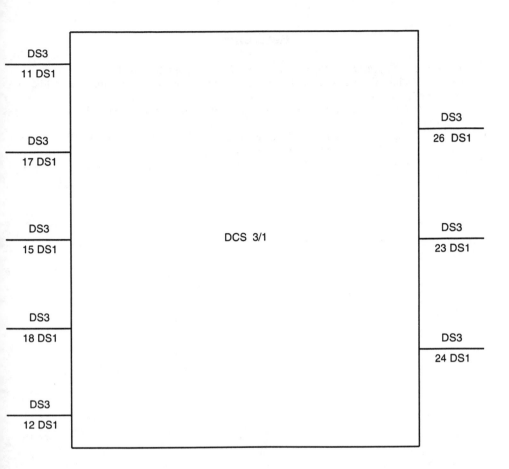

Figure 3.8 DS3 grooming with DCS 3/1.

References

[1] Owen, Frank F.E., *PCM and Digital Transmission Systems*, McGraw-Hill, 1982.

[2] Bellcore, Technical Reference TR-TSY-000009, Asynchronous Digital Multiplexes Requirements and Objectives.

[3] ANSI T1.102-1987, American National Standard for Telecommunications, Digital Hierarchy-Electrical Interfaces.

[4] Bellcore, Technical Reference TR-TSY- 000233, Wideband and Broadband Digital Cross-Connect Systems Generic Requirements and Objectives.

[5] Bellcore, Technical Advisory TA-TSY-000241, Electronic Digital Cross-Connect (EDSX) Frame Requirements and Objectives.

DS0 and Subrate Signals | 4

4.1 SYNCHRONOUS DIGITAL DATA TRANSMISSION

In addition to the large and growing volume of high-speed data transmission at the DS1 and higher rates, there is a great deal of data that uses transmission speeds of 56 Kbps and 64 Kbps, for which the DS0 channel is well suited. There still exists some data traffic at lower speeds, and it is not economical to use a whole DS0 channel for only one of these lower speed streams. For this traffic a multiplexing scheme is employed to combine 2.4-, 4.8-, 9.6-, and 19.2-Kbps streams into one DS0 channel. The volume of this subrate data traffic is declining fast, but the information on the techniques that are used is included, as it is still encountered in some networks.

A point-to-point circuit connects two data customer locations. Multiple customer sites may be connected by a multipoint circuit using multipoint junction units (MJU). In the downstream direction, a control site broadcasts to the remote stations. Upstream, the signals transmitted by the remote sites are combined in the MJUs into a serial bit stream for transmission to the control station. The data customer is responsible for ensuring discipline on the line to avoid collisions of bit streams. All sites on a multipoint circuit operate at the same data-transmission speed, and no direct transmission between remote sites is provided.

In addition to the primary channel, an independent secondary channel capability is available at a bit rate lower than that of the primary channel. Two remote sites on a multipoint circuit may communicate with the control site simultaneously, with one using the primary channel and the other using the secondary channel.

4.2 DS0 DATA CHANNELS

Each DS0 channel in the DS1 frame may be used to carry both primary and secondary channel data. A DS0 channel may carry data from only one source or may carry data from two or more sources multiplexed into one stream. The single-source channel is designated DS0A and a channel carrying data from multiple sources is called DS0B. Error correction for DS0A channels carrying 2.4-, 4.8-, and 9.6-Kbps data is performed within the channel. For 19.2-, 56-, and 64-Kbps DS0A traffic, and for DS0B channels, error correction algorithms are used and a second DS0 channel may be provided for the error-correction function. The sub-rate data streams are encoded so that they—and various added overhead bits—occupy the available bit positions in the DS0 signal.

The zero code suppression (ZCS) is used to overwrite bytes which consist of all zeroes. The ZCS pattern is 00011000. It is used where required in the lower subrate DS0B, 19.2-, 56-, and 64-Kbps channels.

4.3 THE DS0A CHANNEL FORMAT

4.3.1 DS0A – Lower Subrate Speeds

The subrate byte consists of eight bit positions, as shown in Table 4.1. Bit positions 2 through 7 are used for customer data, and bit 1 is set to a one to ensure that in all cases the byte contains at least one 1. The eighth position is used for a network control or control/secondary bit. This 8-bit subrate byte is repeated N times to achieve the 64-Kbps rate of the DS0 channel. The value of N is 5, 10, or 20 for the 9.6-, 4.8-, and 2.4-Kbps rates, respectively.

Table 4.1
DS0A Channel Format – Lower Subrate Data

	Bit Positions							
	1	2	3	4	5	6	7	8
Bit Values	1	D	D	D	D	D	D	C

Notes:
D = Customer data.
C = Network control or control secondary bit.

As an example, the value of N is 20 for the 2.4-Kbps case. As 20 subrate bytes make up a 64-Kbps bit stream, then each of the 20 subrate bytes has a rate of 3.2 Kbps. Each subrate byte contains six data bits and two overhead bits, so the data rate is 0.75 times the byte rate, or 2.4 Kbps.

4.3.2 DS0A – 19.2-Kbps Rate

The format of the 19.2-Kbps DS0A frame is shown in Table 4.2. The 5-byte frame is identified by the framing pattern 01100, which uses bit 1 of each byte. Bits 2 through 7 of bytes 2 and 3 are assigned to the 19.2-Kbps customer-data information or to control codes, and bit 8 of these two bytes is a network control or a control/secondary channel bit. The network control bit 8 is set to a one to indicate that primary channel data are being carried, and bit 8 is set to zero to indicate when control codes are being carried in bits 2 through 7. In bytes 1, 4, and 5, bits 2 through 8 are undefined except when the channel is using special features such as latching loopback, but an all-zeros code is not used.

Table 4.2
DS0A Channel Format – 19.2-Kbps Data

		Bit Positions							
		1	2	3	4	5	6	7	8
Bit Values	Byte 1	0	X	X	X	X	X	X	X
	Byte 2	1	D	D	D	D	D	D	C
	Byte 3	1	D	D	D	D	D	D	C
	Byte 4	0	X	X	X	X	X	X	X
	Byte 5	0	X	X	X	X	X	X	X

Notes:
D = Customer data.
C = Network control or control secondary bit.
X = Not assigned, except that an all-zeros code is not used.

The five bytes, containing 40 bits, that make up the 19.2-Kbps subrate frame operate at the 64-Kbps rate, so each of the subrate bits has a rate of 1.6 Kbps. The 40 bits are made up of 12 data bits and 28 overhead bits, so the data bits operate at a rate of $12 \times 1.6 = 19.2$ Kbps.

4.3.3 DS0A – 56-Kbps Rate

The format is shown in Table 4.3. The first seven bits of each byte contain customer data, and the eighth bit is used for control or secondary-channel data. Because this byte occupies the complete DS0 channel, there is no DS0B format for 56-Kbps data.

Table 4.3
DS0A Channel Format – 56-Kbps Data

	Bit Positions							
	1	*2*	*3*	*4*	*5*	*6*	*7*	*8*
Bit Values	D	D	D	D	D	D	D	C

Notes:
D = Customer data.
C = Network control or control secondary bit.

4.3.4 DS0A – 64-Kbps Rate

As shown in Table 4.4. all eight bits are used for customer data, except when the channel is out of service for maintenance. At those times, bits 1 through 8 are used for network control and status purposes. This byte also occupies the entire DS0 channel, so there is no 64-Kbps DS0B format.

Table 4.4
DS0A Channel Format – 64-Kbps Data

	Bit Positions							
	1	*2*	*3*	*4*	*5*	*6*	*7*	*8*
Bit Values	D	D	D	D	D	D	D	D

Notes:
D = Customer data.

4.4 THE DS0B CHANNEL FORMAT

4.4.1. DS0B – Lower Subrate Speeds

The DS0B channel combines data from two or more sources and may include 9.6-, 4.8-, or 2.4-Kbps data streams. As a result, one DS0B channel contains a number of DS0A channels at one of these subrate speeds. The DS0B channel format for the lower subrate speeds is shown in Table 4.5. The DS0B frame contains *N* bytes, where *N* is equal to the number of DS0A streams it contains.

For 9.6-Kbps traffic, $N = 5$; for 4.8-Kbps data, $N = 10$; and $N = 20$ for the 2.4-Kbps DS0A channel case. Each byte carries the framing bit in position 1, bits 2 through 7 contain customer data and bit 8 carries control bit or control/secondary channel information. The DS0B framing patterns are used to allow the de-multiplexing of the DS0A signals from the DS0B channel at the receiver, and the bit pattern varies among the three channel speeds as detailed in Table 4.5.

Table 4.5
DS0B Channel Format – Lower Subrate Data
I

		Bit Positions							
		1	2	3	4	5	6	7	8
Bit Values	Byte 1	F	D	D	D	D	D	D	C
	Byte 2	F	D	D	D	D	D	D	C
	Byte N	F	D	D	D	D	D	D	C

Notes:
D = Customer data.
C = Network control or control secondary bit.
F = Framing bit.

II

Data Speed	*N*	*Framing Pattern*
9.6 Kbps	5	01100
4.8 Kbps	10	0110010100
2.4 Kbps	20	01100101001110000100

4.4.2 DS0B – 19.2-Kbps Format

The DS0B channel uses a 5-byte format to carry two 19.2-Kbps streams as well as a 9.6-Kbps rate data channel. This is illustrated in Table 4.6. Bit 1 of each byte carries the framing information, and the pattern used is 01100. The six bits 2 through 7 of each byte carry the customer data. In byte 1, the customer data is the subrate 9.6-Kbps channel. Bytes 2 and 3 carry the first 19.2-Kbps stream and bytes 4 and 5 are used for the second 19.2-Kbps channel. In all five bytes, bit 8 is used for control or control/secondary channel bit for the customer data in that byte.

Table 4.6
DS0B Channel Format – 19.2-Kbps Data

		Bit Positions							
		1	2	3	4	5	6	7	8
Bit Values	Byte 1	0	D	D	D	D	D	D	C
	Byte 2	1	E	E	E	E	E	E	C
	Byte 3	1	E	E	E	E	E	E	C
	Byte 4	0	F	F	F	F	F	F	C
	Byte 5	0	F	F	F	F	F	F	C

Notes:
D = Customer data for 9.6-Kbps subrate channel.
E = Customer data for first 19.2-Kbps channel.
F = Customer data for second 19.2-Kbps channel.
C = Network control or control secondary bit.

4.5 CONTROL CODES

Control codes sent in the DS0 channel are used for maintenance and control purposes. The codes show data network status and can be used to activate trouble sectionalization tests.

Table 4.7 lists the control codes used in the synchronous digital data network. The idle code indicates that no customer data is being transmitted and can be used by the customer or by the network to indicate that the DS0 channel is to be split for testing.

Three loopback conditions may be used to isolate a fault condition on a data channel. OCU loopback takes place in the office channel unit at the interface to the local loop and is used to determine if a fault is in the channel or on the loop. Channel loopback occurs at the front end of the customer premises equipment (CPE) and isolates a fault to the loop or CPE. The optional DSU loopback is applied at the customer side of the CPU, for further fault isolation. For these three loopback modes, at an interval after initiation the DS-0 signal from the testing center will change from a steady control byte sequence to an

Table 4.7
Control Codes

Message	Bit Positions							
	1	2	3	4	5	6	7	8
Idle	S	1	1	1	1	1	1	0
OCU Loopback	S	0	1	0	1	0	1	0
Channel Loopback	S	0	1	0	1	0	0	0
DSU Loopback	S	0	1	0	1	1	0	0
Abnormal Station (ASC)	N	0	0	1	1	1	1	0
Mux-out-of-Sync (MOS)	N	0	0	1	1	0	1	0
Unassigned Mux Channel (UMC)	N	0	0	1	1	0	0	0
Test Code	S	0	0	1	1	1	0	0
Test Alert (TA)	S	1	1	0	1	1	0	0
MJU Alert (MA)	S	1	1	1	0	0	1	0
Loopback Enable (LBE)	S	1	0	1	0	1	1	0
Far-End Voice (FEV)	S	1	0	1	1	0	1	0
Transition in Progress (TIP)	S	0	1	1	1	0	1	0
Block Code (BLK)	S	0	0	0	1	0	1	0
Release Code (RLS)	S	1	1	1	1	0	0	0
Zero Suppression (ZCS)	S	0	0	1	1	0	0	0

NOTES:
S = Appropriate subrate framing bit when byte is a 19.2-Kbps DS0A signal
    or part of a DS0B signal.
S = "Don't Care" when byte is any other DS0A signal, except for zero
    code suppression (ZCS).
S = Zero for ZCS when it is a DS0A signal except for 19.2-Kbps rate.
N = "Don't Care."

alternating "N control bytes/N test bytes" sequence. For the OCU and channel loopback modes, the OCU must retain the loopback under the alternating byte condition.

The next four control codes indicate that the channel is out of service for the reason indicated. The following seven codes are used for testing the digital data channel, with *loopback enable* providing a latching loopback feature. The ZCS feature is used to overwrite an all-zeros byte.

In the case of 64-Kbps customer data that uses all eight bits for traffic, the data will replicate the control-code bytes. To avoid problems in 64-Kbps data applications, it may be necessary to restrict customer generation of certain sequences of control-code bytes. This is true only in digital data service. In clear-channel 64-Kbps service, or in an integrated services digital network (ISDN), this restriction does not apply.

DS1 Signal   5

5.1 INTRODUCTION

The first level of the North American digital hierarchy is the DS1, whose basic parameters, established with the introduction of the T1 carrier system, include the signal rate of 1.544 Mbps and the 193-bit frame structure [1]. The message capacity of each frame is 192 bits, which may be divided into 24 DS0 channels of eight bits each. The 193rd bit is designated the frame overhead bit; it is provided to mark the beginning of each frame. At the receiving end of the system, the terminal equipment sees a continuous stream of digits at the 1.544-Mbps rate, and it is the detection of the framing overhead bits that allows the position of each frame to be located. Once this is accomplished, the eight bits comprising each DS0 channel are sent to the correct channel receiver. As the digital network evolved, the applications of the message bits and the techniques used for framing have changed. Various DS1 signals may co-exist in the same network, but they may be incompatible so that they cannot be connected together on an end-to-end basis—even though they all have the common 1.544-Mbps speed.

The attributes of the DS1 signal are the frame format, the line coding, frame synchronization, and clock synchronization. The frame format has evolved from the original T1 framing to the current superframe and extended superframe versions in use today. The second of these attributes complicating interconnection is the line coding used on DS1 transmission links to overcome the problem of long strings of zeros in the bit stream. Two coding formats are in common use, and a network element such as the DCS 1/0 cross-connect system has to have its ports correctly provisioned to send and receive the appropriate line code for each application. This chapter traces the evolution of the DS1 signal and explains the various formats that may be encountered in digital networks.

Framing synchronization requires the framing detector at the receiver to correctly identify the framing bit pattern sent by the transmitter and not lock on to a false framing pattern. If the framing detector loses its synchronization to the correct framing pattern, an out-of-frame (OOF) condition is declared, and normal operation is suspended until framing synchronization is regained.

Clock synchronization refers to the provision of precise clock frequencies at the terminal sites of a system. In a master-slave relationship, the slave site synchronizes its clock to the signal incoming from the master site using a technique called clock recovery. In another network arrangement, all the sites derive their clock frequency from a high-precision clock source, which may be provided at each site or distributed to all the sites from one central source. This is discussed in more detail in Chapter 7, which covers synchronization.

5.2 THE ORIGINAL T1 SYSTEM

The original T1 PCM system, introduced into the Bell System around 1962, used the D1 PCM channel bank as its terminal equipment. Of the eight bits dedicated to each channel, seven were used for encoded speech and the eighth carried signaling information. The framing overhead bits of successive frames were set to produce a 10101010 pattern.

Many of the initial T1 systems were used to reduce the number of voice frequency cable pairs needed for interconnecting telephone offices. Many of these links were short, and analog cable systems had not proved economical in these situations. The T1 system with the D1 channel bank proved to be a good solution for interoffice systems, providing adequate transmission performance for this and other point-to-point applications. It consisted of a PCM channel bank at each terminal and some intermediate regenerators. However, two problems became apparent as the T1 systems were more widely deployed. The first of these was that the 7-bit speech coding was inadequate for providing proper transmission quality when T1 systems were connected in tandem at the voice-frequency level to create long-distance circuits. To prevent the occurrence of this problem, the number of T1 systems in series had to be limited to three, which substantially complicated network provisioning and circuit planning, particularly where more than one telephone company was involved in providing service.

The second difficulty became apparent when voice-frequency tones were used on the channels of the D1 channel banks to align and test the T1 systems. It was discovered that some combinations of these tones caused alternating ones and zeros that were spaced 193 bits apart to appear in one or more of the channel allocations of the DS1 bit stream, and these falsely duplicated the 101010 framing pattern. This problem did not become apparent as long as the system remained locked onto the original correct framing bit sequence. If errors causing framing to be lost occurred in the line signal, the receiver was unable to distinguish between the proper framing bits and the false 101010 patterns caused by the test signals. The resulting loss of frame condition caused traffic to be disrupted on all 24 channels until reframing was achieved. In general, the simple alternating ones and zeros framing pattern was found to be insufficiently robust in the presence of line errors, and framing was lost too frequently.

5.3 THE DS1 IN THE TOLL NETWORK

By the late 1960s, developments in digital technology had reached the point where the feasibility, in the future, of a completely digital telecommunications network was becoming apparent. Not only were transmission systems operating at rates of over 100 Mbps being successfully demonstrated, but digital message switches having digital ports in place of voice frequency (VF) terminations were becoming practicable. The DS1 level would be used for interconnection between network elements within a digital office, so that the DS1 signal as the first level in the hierarchy had to be of toll quality in all respects. A PCM channel bank suitable for the toll network was essential to replace the D1 bank. The loss of frame problem associated with the D1 channel bank could not be tolerated in the network because with multiple DS1 interconnections a loss of frame condition would be propagated through the network. This would cause successive reframing actions in a number of network elements, disabling large numbers of circuits.

5.4 ROBBED-BIT SIGNALING

A toll-quality channel bank, the D2, was introduced to overcome the performance problems of the D1 bank. The D2 introduced a new DS1 format, which used all eight bits of each channel for speech coding in five frames out of six, and in the sixth frame used seven bits for coding and one for signaling. This technique, known as robbed-bit signaling, allowed toll-quality transmission performance to be achieved and removed the restriction on the number of systems that could be connected in tandem. The more limited number of bits available for signaling proved to be sufficient, and this format was retained in subsequent versions of the PCM channel bank.

With the D1 channel bank every frame had the same format, and the framing detection process involved finding the 101010 pattern formed by the successive framing overhead bits. The D2 format was more complicated—not only was it necessary to mark the beginning of each individual frame, but the positions of those frames that used robbed-bit signaling had to be indicated also.

This was accomplished by introducing the superframe format described in the next section.

5.5 SUPERFRAME

5.5.1 Frame Structure

The 2,316-bit superframe consists of 12 successive individual DS1 frames, as shown in Figure 5.1. Because robbed-bit signaling takes place in every sixth frame, it appears in frames 6 and 12 of the superframe.

5.5.2 Frame Overhead

The 12 frame overhead bits of the superframe have to provide both the individual frame indications and the locations of the robbed-bit signaling frames. The overhead bits of the odd-numbered frames form the frame alignment channel and are designated F1 through F6. The corresponding bits in the even-numbered frames form the superframe alignment channel and are designated S1 through S6. The framing overhead bit assignments and the values to which they are set are shown in Table 5.1.

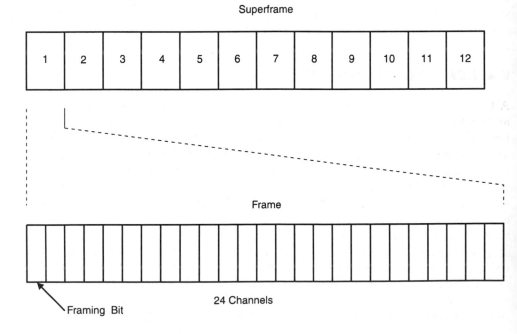

Figure 5.1 Superframe frame structure.

Table 5.1
Superframe Overhead Structure

	F BITS	
	---	---
Frame Number	Frame Alignment Channel	Superframe Alignment Channel
1	F1 = 1	
2		S1 = 0
3	F2 = 0	
4		S2 = 0
5	F3 = 1	
6		S3 = 1
7	F4 = 0	
8		S4 = 1
9	F5 = 1	
10		S5 = 1
11	F6 = 0	
12		S6 = 0

The frame alignment channel is used to locate the beginning of the frames, and the superframe alignment channel is used to locate the beginning of the 12-frame superframe, which also marks the positions of the sixth and twelfth frames. The superframe framing sequence is 100011011100, which is much more complex than the D1 pattern of 101010 and makes false framing a much less frequent occurrence. Fortunately, the integrated circuits that were available for the design of the D2 and subsequent channel banks were much more sophisticated than the components used in the D1, and they made possible the necessary techniques required to implement the superframe formats and frame detection.

5.5.3 System Performance

The superframe format allowed the principal shortcomings of the D1 single-frame DS1 signal to be overcome. The format is more sophisticated than the original single frame, but all of the framing overhead bits are still assigned to framing functions, and none is available for other purposes. No connectivity between the PCM terminals is available apart from the 24 message channels, and no system capacity can be used to assess the system performance in service. Out-of-service performance testing can be carried out using test patterns applied by a DS1 bit-error test set, but this involves taking all 24 channels out of service. It was to overcome these problems that the extended-superframe format was developed.

5.6 EXTENDED SUPERFRAME

5.6.1 General

After the superframe format was established, development continued on framing techniques. Improvements in the precision of the local-site clocks and the introduction of transmission systems with much lower error rates allowed framing synchronization to be carried out using fewer framing bits. The use of a 24-frame format, called the extended superframe, allowed all of the framing requirements to be satisfied using only six of the 24 available framing overhead bits. This made 18 overhead bits available for other purposes, including in-service system performance assessment and the provision of a communication link between the terminals.

5.6.2 Frame Structure

The structure of the DS1 extended superframe is shown in Figure 5.2. The 4,632-bit DS1 extended superframe consists of 24 individual frames, each of which has a total of 193 bits. Of these 193 bits, 192 are available for payload, with one frame overhead bit.

5.6.3 Frame Overhead

The extended superframe includes 24 frame overhead bits, as shown in Table 5.2. These are divided into the extended superframe alignment channel (bits F1 through F6), the cyclic redundancy check (C1 through C6) and the data link (M1 through M12). The functions of each of these is described below. It is the availability of the cyclic redundancy check and the data link capabilities that makes the extended superframe format a great improvement over the superframe.

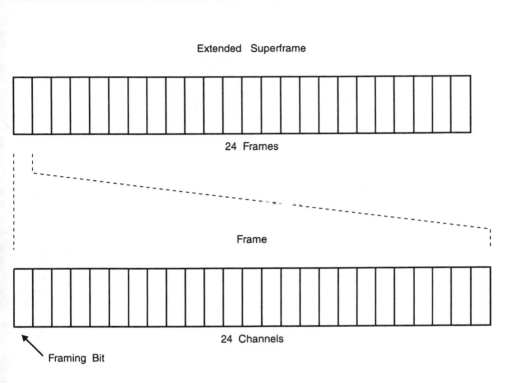

Figure 5.2 Extended superframe frame structure.

Table 5.2
Extended Superframe Overhead Structure

	F BITS		
Frame Number	*Frame Alignment Channel*	*Data Link*	*Cyclic Redundancy Check*
1		M1	
2			C1
3		M2	
4	F1 = 0		
5		M3	
6			C2

Table 5.2 (continued)
Extended Superframe Overhead Structure

| | *F BITS* | | |
| | Frame Alignment Channel | Data Link | Cyclic Redundancy Check |
Frame Number			
7		M4	
8	F2 = 0		
9		M5	
10			C3
11		M6	
12	F3 = 1		
13		M7	
14			C4
15		M8	
16	F4 = 0		
17		M9	
18			C5
19		M10	
20	F5 = 1		
21		M11	
22			C6
23		M12	
24	F6 = 1		

5.6.4 Extended Superframe Alignment Channel

The framing functions of the extended superframe are performed by the six F-bits, which comprise the extended superframe alignment channel. These framing functions are the same as those performed by the F-bit and S-bit patterns in the superframe format, but are in a different pattern so that a channel bank using the extended superframe format will not synchronize with one using the superframe format. More sophisticated technology allows the simpler framing bit pattern to achieve frame synchronization with a minimum of false framing problems.

5.6.5 Cyclic Redundancy Check

The cyclic redundancy check (CRC-6) makes use of the six C-bits to allow in-service DS1 performance monitoring. The basis of this technique is to examine the composition of one extended superframe (ESF) bit stream at the transmitting end, and then examine the same bit stream again at the receiving terminal of the system. Calculations based on the bit-stream composition in each case give the values to be assigned to the C1 through C6 bits. At the transmitter, the C-bit values cannot be calculated until the complete ESF block has been transmitted. These values are then inserted into the C-bit positions of the following ESF block. At the receiver, the C-bit values are calculated when the ESF is received, and the two sets of C-bit values are compared. If no errors have occurred in the transmission path, the C-bit values calculated at the transmitter, and sent in the following ESF, are identical with those calculated at the receiver. The detailed method of calculating the C-bit values is as follows: Before the calculation is carried out, each frame overhead bit is set to 1, leaving all other bit values in the ESF unchanged. The 4,632 bits that make up the ESF constitute the CRC-6 message block (CMB). The check bit sequence C1 through C6 is the remainder after the polynomial representing the CMB has been multiplied by X^6 and divided by the generator polynomial $X^6 + X + 1$. The most significant bit of the remainder is C1, and the least significant bit is C6.

5.6.6 Data Link

The data link bits M1 through M12 appear in the odd-numbered frames of the ESF (Table 5.2). They provide a 4-Kbps data link over a DS1 path, which has two defined applications. The first is as a DS1 data link channel for system messages. The second use is for ZBTSI applications, which are discussed in Section 5.7.4.

5.7 DS1 SIGNAL CODING

5.7.1 General

The discussion of the DS1 signal in the prior sections has concentrated on the various formats of the DS1 bit streams. The DS1 level of the hierarchy appears in the network at interface points between network elements and in line transmission systems [2]. These DS1 line systems are generally used on low-density feeder and distribution routes. In both of these cases, the transmission medium is the metallic balanced cable pair. Successful transmission of digital signals on a metallic pair cable requires the satisfaction of three major conditions:

- The pulse stream must have no average direct current (DC) component, so line coding is used to provide a bipolar signal.
- In order for the pulse stream to be regenerated correctly it must not contain more than 15 consecutive zeros. The timing signal used for regeneration is derived from the incoming DS1 signal, and it is more difficult to extract the timing information if more than 15 zeros occur between pulses.
- The ones density must be not less than 12.5%.

If the signal generated by the transmitter does not satisfy these conditions, some form of line coding is necessary before the DS1 stream is applied to the cable pair.

5.7.2 AMI

In a unipolar DS1 signal, all of the pulses representing ones are of the same polarity—either all positive or all negative—and the pulse stream has a DC component. If a unipolar signal were sent on to the line, the DC component would be removed at the input transformer of the first regenerator it encountered. A bipolar signal, on the other hand, has no DC component, since the positive and negative pulses balance one another. The alternate mark inversion (AMI) technique, illustrated in Figure 5.3, provides conversion from unipolar to bipolar configuration by reversing the polarity of every other pulse. Errors introduced in the transmission path may result in the received signal having successive pulses of the same polarity, which is called a bipolar violation. Detection of bipolar violations may be used to initiate alarms or to trigger line protection systems to switch traffic off the line with the error problem.

An advantage of the bipolar format is that the maximum energy in the pulse train is at a frequency one half of the pulse repetition rate of 1.544 Mbps, and so a bipolar signal has its maximum energy at 772 kHz. A unipolar signal, however, has its maximum energy at 1.544 MHz, equivalent to the 1.544-Mbps pulse rate. Two benefits of operating at the lower frequency are that the loss of the cable is lower and there is less crosstalk coupling between cable pairs. The AMI line coding meets the first of the conditions by removing the DC component of the line signal. The not-more-than-15-consecutive-zeros constraint is met with a PCM channel bank carrying voice traffic, due to the speech and signaling activities on the channels, and the AMI line-coding technique may be used in these applications. With 56-Kbps data, the eighth bit in each can be used to ensure that some ones are included in the bit stream. However, clear channel

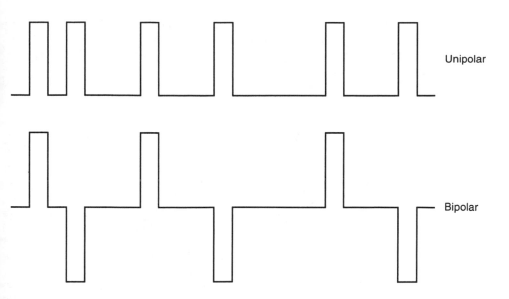

Figure 5.3 Alternate mark inversion.

operation, where 64-Kbps data traffic may be used on any of the channels and signaling is not used, can create conditions where the not-more-than-15-consecutive-zeros rule is not met and AMI coding is not satisfactory.

5.7.3 B8ZS

Line-coding techniques that remove the not-more-than-15-consecutive-zeros constraint provide bit-sequence independence of the DS1 signal and allow clear channel transmission. In the bipolar with eight zero substitution (B8ZS) coding scheme, any sequence of eight zeros is replaced by the bit pattern 00 0VB 0VB. In this pulse train, B represents a normal bipolar pulse and V is a bipolar violation pulse. A bipolar violation occurs when a one has the same polarity as the last preceding one.

When B8ZS signals are detected at the receive terminal decoder, the bipolar violations are removed, and the original eight-zeros sequence is restored. If, however, the B8ZS signal is applied to the receiver of any network element provisioned for AMI signals, the bipolar signal is converted to unipolar form, including the ones added in the B8ZS coding at the transmit end. This introduces errors since eight zeros have been replaced by the sequence 00011011. It is essential that the sending and receiving terminals of the line system are both provisioned for the same line coding—either AMI or B8ZS.

The B8ZS-coded DS1 signals may cause problems on line systems that give alarms or perform protection switching when the receiver detects the presence of bipolar violations. This may restrict the applications of the B8ZS-coding technique. It should also be noted that both AMI and B8ZS line codes are used only on metallic pair cable systems. They are not used when the DS1 signal is applied to higher order multiplexers.

5.7.4 ZBTSI

The zero byte time slot interchange (ZBTSI) technique is used to provide bit-sequence independence to allow clear channel operation of DS1 signals. The ZBTSI coding ensures that there are never more than 15 consecutive zeros, and also that the ones density in the bit stream is at least 12.5%. Unlike B8ZS line coding, ZBTSI is an end-to-end encoding and decoding technique, which operates on the data in the bit stream prior to any line coding. It passes through the network from DS1 terminal to DS1 terminal, and uses some of the DS1 overhead bits to signal that ZBTSI coding is being used.

Referring to Table 5.2, there are 12 M-bits in the ESF format. The odd-numbered data channel bits are used for ZBTSI encoding flags at a 2-Kbps rate, and they are then referred to as Z-bits. The even-numbered data channel bits constitute a 2-Kbps maintenance channel. The remaining ESF framing bits retain their normal function. The ZBTSI technique regards each DS1 frame as 24 eight-bit channels plus a framing bit in the first bit position. The ZBTSI encoder operates on blocks of 96 contiguous 8-bit channels (octets) and numbers them from 1 through 96. Each 96-octet block is immediately preceded by a ZBTSI encoding Z-bit. The framing bits between the Z-bits are not involved in the ZBTSI encoding process.

Prior to entering the encoder, the PCM data is scrambled in a device synchronized to the ESF framing sequence. Each octet is examined in the encoder together with the two adjacent octets. An all-zero octet must be processed when it combines with the adjacent octets to form a string of 15 or more zeros, or an insufficient number of ones is found in the bit sequence. If no encoding is required in the 96-octet group, the flag bit is transmitted, followed immediately by octet 96 and then octets 1 through 95 in sequence.

When 15 or more zeros or an insufficient ones density is found in an octet, the Z-bit for that group of 96 octets is changed from a one to a zero. The location number of the violating octet is determined, and the number is encoded into a 7-digit binary pattern. The data in octet position 96 are stored separately, and the 7-bit code of the violating octet number is placed in the octet-96 location. If there is only one violating octet, the first digit in octet position 96 is set to a one. The original data from octet 96 are placed in the location of the violating octet. If more than one violating octet is found, the first bit in the octet-96 position is set to zero, and the remaining seven digits are the location of the first violating octet. The 7-bit location of the second violating octet is stored in the location of the first violating octet, with the first digit being a one if there are more violating octets or a zero if there are no more after the second. This process is continued until all of the violating octets have been identified, and the original data from octet 96 is then placed in the location of the last violating octet.

The F-bits that were excluded from the ZBTSI encoder are buffered to equalize the delay experienced by the information bits, so that the original superframe structure is recovered at the decoder. Where the ZBTSI encoding takes place at a DS1 terminal, the CRC-6 check detailed in Section 5.6.5 is performed before the data enters the encoder. The original bit sequence is restored at the decoder at the receive end of the link.

The ZBTSI technique is complicated and involves a great deal of circuitry to handle the encoding, decoding, and buffering functions. In addition, the buffering process introduces substantial delay in the transmission of the DS1 information.

References

[1] ANSI T1.107-1990, American National Standard for Telecommunications, Digital Hierarchy-Formats Specifications.

[2] Owen, Frank F.E., *PCM and Digital Transmission Systems*, McGraw-Hill, 1982.

DS3 Signal — 6

6.1 INTRODUCTION

The DS3 signal is the third level in the North American digital hierarchy, and has a nominal bit rate of 44.736 Mbps [1]. It is used as the interface between digital transmission systems and the network elements on which they terminate at terminal and junction sites. The DS3 frame structure was originally designed to accommodate the traffic and overhead bits resulting from the asynchronous multiplexing of DS1 and DS2 signals into the DS3 format. In this format, the overhead bit capacity available for performance monitoring and other maintenance functions is limited. As long as the traffic on the network was mainly digitized speech this did not cause too much concern, but the ever-increasing volume of data traffic made the need for these capabilities much more urgent. First SYNTRAN and then C-bit parity DS3 formats have been introduced, which enable expanded performance monitoring and other maintenance functions to be provided. All use the same DS3 frame structure. The frame structure, along with the format and applications of these DS3 signals, are described in the following sections.

6.2 DS3 FRAME STRUCTURE

The frame structure is shown in Figure 6.1. The M-frame is 4,760 bits long and consists of seven M-subframes of 680 bits each. The M-subframe includes eight frame overhead bits, each of which is followed by 84 payload bits.

6.3 FRAME OVERHEAD STRUCTURE

The M-frame includes 56 frame overhead bits as shown in Table 6.1. These are divided into the M-frame alignment channel (M1 through M3), M-subframe

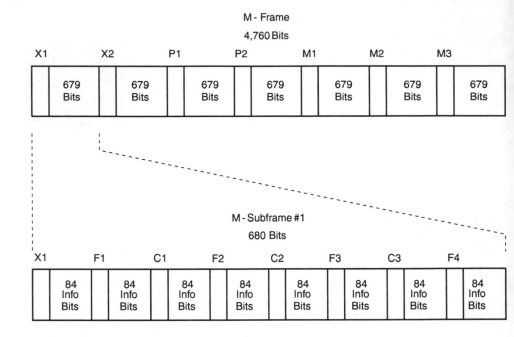

Figure 6.1 DS3 frame structure.

alignment channel (F1 through F4), P-bit channel (P1 and P2), X-bit channel (X1 and X2), and the C-bit channel (C1 through C3). The function of each of these is described below.

Table 6.1
DS3 Frame Overhead Structure

M-Subframe Number	M-Frame Alignment Channel	M-Subframe Alignment Channel	X-Bit Channel	P-Bit Channel	C-Bit Channel
1			X1		
		F1 = 1			
					C1
		F2 = 0			
					C2

Table 6.1 (continued)
DS3 Frame Overhead Structure

M-Subframe Number	M-Frame Alignment Channel	M-Subframe Alignment Channel	X-Bit Channel	P-Bit Channel	C-Bit Channel
		F3 = 0			
					C3
		F4 = 1			
2			X2		
		F1 = 1			
					C1
		F2 = 0			
					C2
		F3 = 0			
					C3
		F4 = 1			
3				P1	
		F1 = 1			
					C1
		F2 = 0			
					C2
		F3 = 0			
					C3
		F4 = 1			
4				P2	
		F1 = 1			
					C1
		F2 = 0			
					C2
		F3 = 0			
					C3
		F4 = 1			

Table 6.1 (continued)
DS3 Frame Overhead Structure

M-Subframe Number	M-Frame Alignment Channel	M-Subframe Alignment Channel	X-Bit Channel	P-Bit Channel	C-Bit Channel
5	M1 = 0				
		F1 = 1			
					C1
		F2 = 0			
					C2
		F3 = 0			
					C3
		F4 = 1			
6	M2 = 1				
		F1 = 1			
					C1
		F2 = 0			
					C2
		F3 = 0			
					C3
		F4 = 1			
7	M3 = 0				
		F1 = 1			
					C1
		F2 = 0			
					C2
		F3 = 0			
					C3
		F4 = 1			

Note: In the SYNTRAN DS3 format, the C-bits are numbered C1–C21.

6.3.1 M-Frame Alignment Channel

The M-frame alignment channel serves to locate the seven subframes.

6.3.2   M-Subframe Alignment Channel

The M-subframe alignment channel is used to identify the positions of all of the frame overhead bits.

6.3.3. P-Bit Channel

The P-bit channel is assigned to performance-monitoring functions. It is provisioned so that P1 and P2 are both either set to one or set to zero. They provide parity information about the previous M-frame, but since they are reset at each line terminal they cannot carry parity information about the complete DS3 path.

6.3.4 X-Bit Channel

The X-bit channel is assigned to alarm purposes. It is provisioned so that X1 and X2 are both either set to one or set to zero.

6.3.5 C-Bit Channel

The C-bit channel bits are used for different purposes in the DS3 applications. However, the first C-bit in M-subframe 1 is used as the application identification channel (AIC), which identifies the specific application of the DS3 frame—M23, SYNTRAN, or C-bit parity. The AIC of the M23 application is a series of random ones and zeros. In the case of SYNTRAN it is a repeating 100100... pattern, whereas in the C-bit parity application the bit is always set to 1.

6.4 M23 APPLICATION

This is the original and still the most widely used DS3 format. Four 1.544-Mbps DS1 signals are multiplexed to form one DS2 signal, as described in the ANSI standards [1]. Then seven DS2 signals at a nominal rate of 6.312 Mbps are interleaved to form the DS3 signal of 44.736 Mbps. Because each DS2 is asynchronous, the bits contributed by the DS2 streams are insufficient. Stuffing is always needed to increase the total number of bits, including the framing overhead bits, to the DS3 rate. Increasing the number of bits is called positive stuffing. The stuff bits are always removed at the receiver before the bit stream is demultiplexed back to the DS2 level.

One time slot is provided in each M-subframe for inserting a positive stuff bit if it is needed. The C-bits of an M-subframe are used to indicate whether or not that subframe contains a stuff bit. If there is stuffing, the C1, C2, and C3 bits are all set to one, and if no stuffing is present these bits are all set to zero. At the receiver, a majority vote is taken to decide whether ones or zeros were transmitted. The majority vote, based on two out of three, is required since an error in this stuffing bit control would result in a loss of framing in the DS3 stream.

The use of a two-step multiplex technique to get from DS1 to DS3 makes the DS2 level of multiplexing accessible. An M1-2 multiplexer performs the first stage, and the M2-3, the second stage. DS2 signals from other transmission systems can be connected directly to the M2-3 multiplexer, providing another level of flexibility in facility planning. The DS2 signal has no performance-monitoring feature and includes only an X-bit that is used for alarm purposes. A signal at the 6.312-Mbps DS2 rate, but without framing or any overhead bits, can be connected to the input of an M2-3 multiplexer. However, the M2-3 must be provisioned to accept only a DS2 signal with the correct format to avoid the generation of an unallowable DS3.

When this technique was being developed, it was assumed that T2 transmission systems operating at 6.312 Mbps would be widely deployed. However, it was found that the attenuation of the balanced pair cables in the existing networks, as well as crosstalk coupling between pairs, did not allow systems operating at the T2 rate to achieve the required performance standards. Special cables were designed with superior characteristics, but proved too expensive to allow T2 systems to be cost-competitive in network applications. By 1970, digital coaxial cable systems had evolved to the point where speeds of over 100 Mbps were attainable using the small bore coaxial cable already widely deployed for analog-system use in many countries. As a result, the T2 system was never widely deployed, and access to the DS2 level in the digital multiplexing equipment became unnecessary. There are, of course, very large quantities of multiplexers in service using the M23 DS3 format, and they are not compatible on an end-to-end basis with the SYNTRAN or C-bit parity formats.

6.5 SYNTRAN APPLICATION

This technique is used to produce a synchronous DS3 signal. It multiplexes 28 DS1 streams directly to the DS3 rate, without the intermediate DS2 level being involved. Since no stuffing is employed, the C-bits are not needed for stuffing indicators and are available for other purposes. Features included in the SYNTRAN DS3 include performance monitoring, far-end block error (FEBE) indication, and an alarm and control channel.

The SYNTRAN DS3 technique was fully developed, and equipment was becoming available, when the SONET system plans were announced. Since SONET provided all the features that SYNTRAN offered, beside allowing very high bit rates on optical-fiber cables on a fully synchronous basis, interest in SYNTRAN died down and little deployment took place. It is possible that some SYNTRAN DS3 facilities may be encountered, and care has to be taken not to cross-connect them with any other type of DS3 facility.

6.6 C-BIT PARITY APPLICATION

This application uses a modified version of the M23 multiplexing scheme to make the C-bits available for performance monitoring and other features. In the first stage of multiplexing, four DS1 signals are interleaved into a pseudo-DS2 signal at a nominal bit rate of 6.3063 Mbps. The second stage multiplexer interleaves seven of these pseudo-DS2 signals as described in Section 6.4, except that in this case bit stuffing takes place in every M-subframe. Because the C-bits are not needed as stuffing indicators, they are available for other purposes. The combination of the seven pseudo-DS2 signals with the DS3-level stuffing bits and the 56-frame overhead bits results in the DS3 signal at the nominal rate of 44.736 Mbps.

The DS3 M-frame includes 21 C-bits, and the assignments used in the C-bit parity application are explained in the following sections. The second C-bit in M-subframe 1 remains reserved for future applications. The three C-bits in each of M-subframes 2, 6, and 7 are unused and are set to one.

6.6.1 Application Identification Channel

As detailed in Section 6.3.5, the first C-bit in M-subframe 1 is used for an application identification channel (AIC). In the C-bit parity case, this bit is set to one.

6.6.2 Far-End Alarm and Control Channel

The third C-bit in M-subframe 1 is used for a far-end alarm and control channel (FEAC). The signal is used to send alarm and status messages back to the near end from the far end and to control DS1 and DS3 loopbacks at the far-end terminal. A list of code words has been established for the alarm reporting and control functions, as detailed in the ANSI standards documents [1].

6.6.3 DS3 Path Parity Bits

In M-subframe 3, the three C-bits are named CP-bits and are assigned to carry parity information relating to the DS3 path. At the transmitting end of the system, the three CP-bits are set to the same value as the P-bits—either all ones or all zeros—and the setting is decided by computing the parity based on the contents of the previous M-frame. Since the CP-bits are not reset as they pass through the network, a majority vote of the received CP-bits recovers the transmitted CP-bit value. This is compared with the computed parity of the previous M-frame as received, and a discrepancy shows that an error has occurred during the passage of the M-frame over the network.

This is not a quantitative indication of parity bit error rate (BER) because multiple parity errors tend to cancel each other out. This tends to occur more often with high BER values. Under low BER conditions, however, the C-bit parity technique is a valuable means of DS3 facility performance monitoring. By storing the parity bit errors derived from the C-bit parity indications over a period of time—for example, 15 minutes or one hour—performance data about the in-service DS3 performance of a system may be generated. Examples include:

- *Number of errored seconds, type A* – defined as a second containing one parity error;
- *Number of errored seconds, type B* – defined as a second containing between 2 and 44 parity errors;
- *Number of severely errored seconds* – defined as a second containing more than 44 parity errors.

6.6.4   FEBE Function

The three C-bits in M-subframe 4 are assigned to the FEBE function. When an error is detected at the receive end of a system, the FEBE bits are used to send a message to the transmitting end that an error has been received. The FEBE bits may be set to any combination of ones and zeros to indicate an error, except that 111 is reserved to show that no error has been detected. The FEBE error indications may be stored and used to determine errored seconds and severely errored seconds in exactly the same manner as is used for the CP-bits.

6.6.5   Path Maintenance Data Link

The three C-bits in M-subframe 5 are designated for use as a 28.2-Kbps terminal-to-terminal path-maintenance data link. A variety of operations messages relating to the DS3 facility may be carried on this channel.

References

[1]   ANSI T1.107-1990, American National Standard for Telecommunications, Digital Hierarchy-Formats Specifications.

Synchronization 7

7.1 INTRODUCTION

The basic purpose of synchronization in a telecommunications network is to ensure that the traffic that enters the network is delivered to its destination with the minimum performance degradation. The synchronization process in analog and digital networks is discussed below, and the remainder of the chapter deals with synchronization of digital networks in more detail.

7.2 ANALOG SYSTEM SYNCHRONIZATION

Analog long-distance systems use single sideband, suppressed carrier frequency division multiplexing (FDM), where each voice frequency (VF) channel is modulated into its own frequency band for transmission over the path and is demodulated back to VF at the receiving end. If the carrier frequency used for modulation at the sending end differs from the frequency used for demodulation at the receiving end, the VF traffic is shifted in frequency by the difference between the two carriers. A maximum difference of two Hertz is allowed between the input VF signal and the output signal under the CCITT recommendations.

Speech may be shifted by several Hertz before any noticeable distortion is audible. However, data traffic carried on VF tones is sensitive to the displacement, and transmission may suffer too high an error rate unless the carrier frequencies are more tightly controlled to reduce the difference. The solution is to send a synchronization tone derived from the carrier generator from one end of the system and lock the frequency of the carrier generator at the other end to the synchronization tone. This ensures that there is no frequency difference between the transmit and receive carriers and allows data transmission with an acceptable error rate. A complete network may be synchronized by sending the synchronization tone from one site and carrying it on the transmission links to all of the other network sites.

7.3 DIGITAL SYSTEM SYNCHRONIZATION

7.3.1 Levels of Synchronization

There are three levels of synchronization of digital systems: bit, frame, and network.

The bits are generated at the sending end of a link, and the receiver must determine whether a one or a zero was sent in each bit slot. Unless the clock that controls the sending of the bits is synchronized with the clock that controls the detection of the bits at the receiver, the bits cannot be correctly identified and error-free transmission cannot be achieved.

The DS1, DS2, and DS3 signals have framing bits arranged in a pattern, which must be identified correctly at the receive end of the system so that the beginning and end of blocks of bits can be found. This allows the data bits of the DS1, DS2, and DS3 streams to be allocated to the correct time slots and routed through the multiplex equipment to their proper destinations. In those cases where the DS1 superframe format is used, frames 6 and 12 must then be identified, in order for the robbed-bit signaling information that they carry to be located. Where the DS1 extended superframe format is employed, frames 6, 12, 18, and 24 must be identified to allow the signaling bits to be located.

Network synchronization ensures that both national and international networks may be interconnected, and that high-speed data, facsimile, and digital video services may be carried over interconnected networks without serious impairment. The timing accuracy of every clock used in the networks must adhere to the recognized standards if error-free performance is to be attained. The source of a synchronization problem is difficult to locate within a single network and far more difficult to locate when interconnected networks are involved.

7.3.2 Timing Methods for a Digital Link

The simplest example of a digital system is a point-to-point link. The transmit clock may operate in any of these modes:

- Free running;
- External timing;
- Loop timing.

The receive timing is recovered from the incoming line signal.

The free-running mode is illustrated in Figure 7.1. A local clock source is used at both of the terminal sites and provides timing for the transmitting functions at the site. The timing accuracy of the clocks must be sufficient enough to allow the bit error rate objectives of the link to be met.

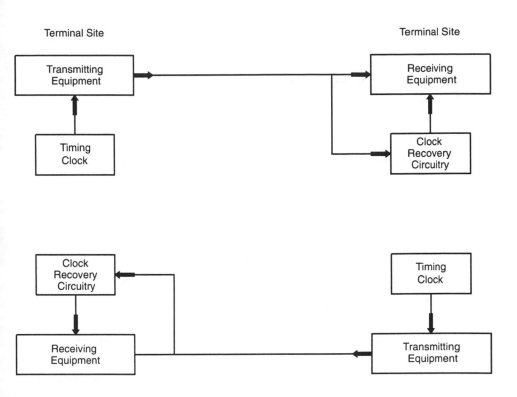

Figure 7.1 Free-running timing.

The external-timing mode is shown in Figure 7.2. A precision timing signal is used to synchronize the transmitting clocks at both terminals.

Loop timing is shown in Figure 7.3. A local clock is used at one terminal site, and the timing information extracted from the received signal at the other terminal is used to time the signals transmitted back to the first terminal site. Loop timing is used on end links where the link terminates in a PCM channel bank or other customer-premises equipment.

7.3.3 Network Synchronization

A digital network consists of a number of node sites interconnected by transmission links. Traffic entering the network may pass through several nodes, including various network elements, before reaching its destination. Ideally, the timing used at each node is at an identical rate, and the network is completely synchronized. In practice, there are small differences between the clock rates at the nodes, so the speed of an incoming pulse stream is slightly faster or slower than the timing rate of the clock at the receiving node. The received pulses are fed into a buffer store, which can accommodate at least two complete DS1 frames, and then read out using the receive-terminal clock. This buffering can

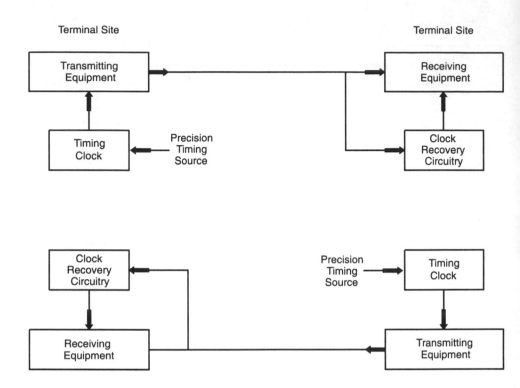

Figure 7.2 External timing.

allow small variations in timing differences to be smoothed out, but when the buffer store becomes empty or overflows, an entire block of 193 bits is repeated or deleted. This event is called a controlled slip.

The repetition or deletion of a DS1 frame in a controlled slip causes a DS1 framing realignment. As mentioned in Section 7.3.1. the superframe (SF) format requires frames 6 and 12 to be identified, and the extended superframe (ESF) format requires frames 6, 12, 18, and 24 to be located, as they carry the robbed bit signaling information. In the event of a controlled slip, the SF or ESF framing has to be realigned to adjust for the deleted or added frame.

7.4 CONTROLLED SLIPS

7.4.1 Effect of Slips

The effect of the controlled slips discussed in Section 7.3.3 depends on the service being carried. Voiceband data suffer strings of errors if even one slip occurs, resulting in the need to retransmit the original data. Digital video is

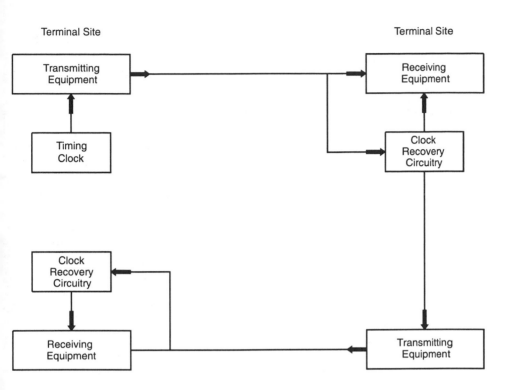

Figure 7.3 Loop timing.

is affected by even one slip, and the amount of distortion suffered by the picture depends on the method of encoding used in the analog-to-digital conversion process.

A controlled slip is a transient phenomenon that may occur at any of a number of locations along the path that carries the service through the network. This makes tracing the source of the problem very difficult. Impairments may be reported by the customer as an excessive bit-error-rate condition, but since the errors are not due to a transmission problem, attempts to trace the cause of the errors to transmission system performance result in "no trouble found" reports. Since any portion of a network must be capable of carrying any and all types of digital traffic, the occurrence of slips must be closely controlled.

7.4.2 Slip-Rate Objectives

For international service, the slip-rate objective has been set by the CCITT at not more than one slip every five hours measured over a 24-hour period. Networks are engineered to meet or exceed this objective, and it is desirable to operate slip-free under normal conditions.

To achieve the slip-rate objective for international connections, the maximum long term frequency inaccuracy of a primary reference source clock is 10^{-11}. Requirements for network clocks are discussed in the next section.

7.5 NETWORK CLOCK HIERARCHY

The slip-rate objectives could be met by providing high-cost primary reference source clocks at every site on a network. Until recently, this would have been prohibitively expensive, but the cost of these clocks has now been drastically reduced. The synchronization method that has been adopted is to send a timing-reference signal from the primary reference clocks to secondary sites, where the local clock locks on to the primary-reference signal and, in turn, sends the reference signal on to other less critical sites. The clocks are arranged in a hierarchical master-slave chain from stratum 1 to stratum 4 [1,2], with the stratum 1 primary-reference clocks having the highest frequency stability. Under normal conditions, all of the clocks are locked to timing references traceable to the stratum 1 clocks, and the slip-rate objective is met in all parts of the network. In the hierarchical arrangement, the clock in a lower stratum can always lock on to a reference signal from a higher stratum clock. However, a higher stratum clock has a narrower pull-in range than a lower stratum clock and will not lock on to a signal from the latter. The accuracy and pull-in ranges of the clocks are detailed in the next section.

To aid in the reliability of the synchronization system, each clock receives primary and secondary timing signals and will lock on to the secondary source when the primary source fails. If both sources fail, the clock has sufficient stability to maintain reasonable performance for a limited period until the reference source is restored. The SONET standards will include a messaging technique, which will allow a terminal site to identify the derivation of the timing source.

In asynchronous networks the performance objectives have been satisfied by providing stratum 2 clocks at toll switching sites, stratum 3 clocks at local switching sites, and stratum 4 clocks with channel banks and digital PBX equipment. The timing of SONET networks, however, is expected to require the deployment of primary-reference sources with stratum 1 performance at many of the major sites on the network, and clocks with at least stratum 2 performance at other major locations.

7.6 STABILITY OF STRATUM 2, 3, AND 4 CLOCKS

Three elements determine the stability of stratum 2, 3, and 4 clocks. These are the minimum accuracy, the holdover requirements, and the pull-in range of the clocks.

7.6.1 Minimum Accuracy

Under normal conditions these clocks are locked to a reference signal. The accuracy requirements for each clock refer to the clock's stability in the free-running mode when no reference signal is applied. The minimum accuracy specifications for the clocks are:

(a) Stratum 2: $\pm$ 1.6×10^{-8}
(b) Stratum 3E: $\pm$ 4.6×10^{-6}
(c) Stratum 3: $\pm$ 4.6×10^{-6}
(d) Stratum 4: $\pm$ 3.2×10^{-5}

The stratum 3E (enhanced) category has the same pull-in/hold-in requirements as the stratum 3 clock, which makes the two compatible. The stratum 3E clock has significantly tighter filtering of short-term instability, long-term accuracy, and holdover requirements. The stratum 3E is expected to be the clock of preference over the stratum 3 for SONET synchronization.

7.6.2 Holdover Requirements

When primary and secondary reference-timing signals are available to synchronize a clock, there should be no adverse effects if the primary-source signal is lost. The complete loss of both the primary and secondary sources of the reference signal causes a stratum 4 clock to pass into the free-running mode. For stratum 2 and 3 clocks, a holdover condition is specified.

(a) Stratum 2 clocks in holdover mode are allowed a frequency accuracy of $\pm 1 \times 10^{-10}$ per day. Under these conditions, some circuits may experience a single slip during the first 24 hours of reference failure.
(b) Stratum 3E clocks in holdover mode are allowed a frequency accuracy of $\pm 1 \times 10^{-8}$ for the first 24 hours.
(c) Stratum 3 clocks can have degraded accuracy in holdover mode, but must not cause more than 255 slips on any circuit in the first 24 hours.

7.6.3 Pull-In Range

The pull-in range of a clock is defined as the maximum range of frequencies for which the clock will lock onto the input synchronization signal. The pull-in ranges are:

(a) Stratum 2: $\pm$ 1.6×10^{-8}
(b) Stratum 3E: $\pm$ 4.6×10^{-6}
(c) Stratum 3: $\pm$ 4.6×10^{-6}
(d) Stratum 4: $\pm$ 3.2×10^{-5}

7.7 PRIMARY REFERENCE SOURCES

7.7.1 Available Sources

Three possible sources are available for the primary-reference frequency for networks operating in the U.S.:

- LORAN-C;
- Global positioning satellite (GPS);
- Atomic clocks.

7.7.2 LORAN-C

The primary purpose of the long-range navigation (LORAN-C) [3] system is for navigation. Originally provided for shipping in U.S. coastal waters and in the Great Lakes, it now covers most of the Northern Hemisphere, Australia, and New Zealand. Its precision as a navigational system is dependent on the frequency accuracy of its transmitted signals, and this leads to its secondary role as a provider of high-precision frequency sources. The accuracy is maintained by comparing three atomic clock standards at each LORAN-C transmitting site. In addition, the frequency is corrected at each of these sites to match the frequency of the international time standard. The frequency accuracy of the LORAN-C signals approaches 1×10^{-13}.

7.7.3 Global Positioning Satellite

The global positioning satellite (GPS) network is a space-based system developed by the U.S. Department of Defense to provide precision position, velocity, and time of day on a continuous, all weather, worldwide, 24-hour basis. The system uses 24 satellites, including three active spares. The satellites orbit at 10,900 nautical miles (20,200 km), and radiate spread-spectrum signals at two frequencies, L1 and L2, which are 1,572.42 MHz and 1,227.6 MHz, respectively. Each satellite uses redundant atomic standard clocks: two Cesium beam and two Rubidium.

The GPS network is under the control of the U.S. government, and its availability for precision-timing purposes might be limited by strategic considerations. Primary-reference sources using secondary atomic clocks can be synchronized to within a few parts in 10^{-13} using the signal from the GPS system.

7.7.4 Atomic Clocks

Stand-alone atomic clocks provide a frequency accuracy of better than 1×10^{-11}, which allows the basic slip-rate objective to be met. However, unless these sources can be compared with the international standards and corrected to conform with them, they cannot provide network performance comparable with the use of the LORAN-C or GPS synchronization methods.

7.8 DISTRIBUTION OF TIMING SIGNALS

Primary-reference sources, traceable to international standards, are located at toll switch sites, which also include many of the cross-connect systems. Distribution of the timing within the sites employs DS1 rate signals. These DS1 signals are also used to synchronize the transmission systems that have equipment at the site and to distribute the timing signals to other sites on the network. By this means a hierarchical synchronization network is established, because a higher level stratum clock passes timing to an equal or lower level stratum clock location, in a master-slave relationship. Again, any difference between the clock rates at the sites at the ends of a link results in controlled slips.

Due to the complex configurations that SONET networks will have when rings are introduced, it is anticipated that more sites will be equipped with primary-reference sources than was the case in asynchronous networks. This is necessary to ensure that the synchronous nature of these networks is maintained, so high-speed data and other new services may be carried with minimum service degradation. Network availability in the SONET environment may be dependent on maintaining proper synchronization. However, the hierarchical clock structure is still employed, and timing signals are sent from higher precision clock sites to those with lower stratum clocks.

A major feature of the SONET technique is that a DS1 or DS3 asynchronous signal is multiplexed into the synchronous payload envelope (SPE) with the use of pointers. These pointers are used to identify the location of the information payload within the SONET frame, and to allow the DS1 and DS3 signals to float within the SPE. The timing differences encountered in SONET networks then result in pointer adjustments, which replace the controlled slips used in asynchronous networks. However, pointer adjustments introduce jitter and make the DS1 signal unsuitable for carrying the synchronization clock in SONET networks. Instead, the OC-N line rate signals are designated as the means of carrying timing information between sites.

References

[1]   ANSI T1.101.1987, Synchronization Interface Standards for Digital Networks and ANSI T1.101-199x (T1X1.3/90-026R7), Revision of T1.101-1987, Synchronization Interface Standard, August 1992.

[2]   Abate, John E., et al, "AT&T's New Approach to the Synchronization of Telecommunication Networks," *IEEE Communications Magazine*, April 1989.

[3]   Loran-C User Handbook, COMDTPUB P16562.6, Superintendent of Documents, U.S. Government Printing Office, Washington, D.C., 1992.

DCS 1/0 System   8

8.1 INTRODUCTION

The digital cross-connect system 1/0 (DCS 1/0) is a centrally controlled network element that provides 64-Kbps DS0 channel cross connection and test access for digital signals that terminate at the DS1 rate. It can electronically cross-connect any DS0 in any DS1 to any available DS0 slot in another DS1 in a nonblocking manner. In addition, the DCS 1/0 also provides DS1, bundled DS0, and subrate signals. Its versatility and cost-effectiveness have provided the impetus for its deployment into various private and public networks. In particular, it has become the workhorse to provide private-line service. The DCS 1/0 utilizes a nonblocking switching matrix to provide dedicated cross connections of digital signals. The different DCS 1/0 systems available that conform to the Bellcore Publication 43801 can transmit signals to one another on an end-to-end basis regardless of which vendor supplied the equipment.

The average transmission delay through the DCS 1/0 was established by Bellcore in the Technical Reference TR-TSY-000170 [1] to be no greater than four 125-microsecond frames and still meet the requirements for jitter tolerance, frame synchronization, and slip control. The maximum delay was set at 0.7 ms.

8.2 DESCRIPTION OF THE DCS 1/0

The DCS 1/0 system can be divided into four basic subsystems consisting of the following:

- Administrative subsystem;
- Switch matrix subsystem;
- Input/output interfaces;
- Synchronization and timing subsystem.

The block schematic is shown in Figure 8.1.

8.3 ADMINISTRATIVE SUBSYSTEM

The administrative subsystem contains the intelligence of the system. It provides diagnostics, provisioning, alarms, communications, and overall control. The microprocessor-based administrative subsystem has both volatile and nonvolatile memories. The working memory is usually in the form of volatile random access memory (RAM). The backup nonvolatile memory normally consists of EPROMs or a hard disk, and in some systems even more backup is provided in the form of a streaming tape for the hard disk. Floppy drives and tape drives are commonly used to load software into the DCS 1/0.

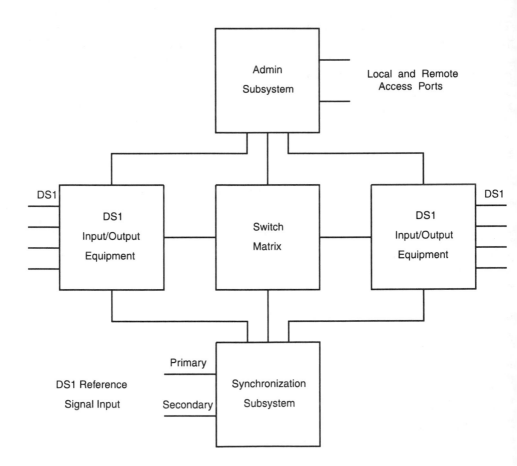

Figure 8.1 DCS 1/0 system block schematic.

8.4. SWITCH MATRIX SUBSYSTEM

The switch matrix subsystem is the internal interface that cross-connects the facilities terminating on the DCS 1/0 system. The cross connection is performed under software command, which allows it to be done remotely. There are two kinds of switching matrices: a time switch called time slot interchange (TSI) and the space-division switch.

Basically, a TSI operates by writing data into and reading data out of a digital memory in the proper sequence under software control. In the process, the digital data of a selected time slot is placed in the desired time slot.

Conceptually, a space switch can be a simple switching structure in the form of a rectangular array of cross-points to connect any one of N inlets to any one of M outlets. To increase the efficiency of the space switch, stages can be cascaded in series. Since the inlet and outlet signals are at the DS1 rate and the switch has to cross-connect DS0 elements, space switching is not applicable to the DCS 1/0 system, and a TSI matrix is used.

8.5 INPUT/OUTPUT INTERFACE SUBSYSTEM

The DS1 input/output (I/O) interface equipment is the physical and electrical interface to the telephone transmission lines, functioning under the control of the administrative subsystem.

The I/O interface ports terminate the incoming digital facilities and pass the signals from the facilities to the switching matrix, and from the switching matrix back to another I/O port and to the outgoing facilities.

8.6 SYNCHRONIZATION AND TIMING SUBSYSTEM

The synchronization subsystem provides a stratum 3 clock in the synchronization hierarchy in order to conform to the network plan for digital synchronization. This synchronization requires the identification and location of the framing pattern in the terminating digital equipment. An elastic store is usually provided at a switching network element such as the DCS 1/0 to take care of small differences in bit clock frequency. This elastic store is often two full frames deep to allow it to operate in the middle to smooth out any clock jitter and to provide controlled frame slips.

Synchronization of the DCS 1/0 to the network clock is accomplished by locking the 1/0 internal clock to a reference signal traceable to a stratum one precision source. Usually, two reference signals are used; one is designated the primary source and the other, the secondary. In the free-running mode, the DCS 1/0 provides a stratum 3 clock. There are various ways a DCS 1/0 can be locked to the network clock, as discussed in Chapter 7.

The synchronization subsystem usually provides:

- Redundant clock reference input lines;
- Phase-lock looped oscillators for frequency control and synchronization;
- Redundant clock monitoring, reporting, and switching;
- Distribution system for the synchronization signals.

8.7 TRADITIONAL DCS 1/0 FUNCTIONS

This section describes some important traditional DCS 1/0 functional characteristics. There are fundamental characteristics shared by all DCS 1/0 systems especially if they are compliant to the Bellcore recommendations. Nonetheless, variations exist from one equipment manufacturer to another. Some of the more common functional characteristics are described here.

8.7.1 DS1 Compatibility

Due to the evolution of the digital channel bank, three channel number assignment sequences were developed over the years. In Table 8.1, three channel assignment sequences are shown. The D1D channel number assignment was developed first. It was followed by the D2 and then the D3/D4. DCS 1/0 compatibility with the different channel assignment sequences is assured when each of the DS1 signals is properly specified during the cross-connection process.

Table 8.1
Channel Number/Time Slot Assignments

Time Slot	Channel Number Assignments		
	D3/D4 Channel Bank	D1D Channel Bank	D2 Channel Bank
1	1	1	12
2	2	13	13
3	3	2	1
4	4	14	17
5	5	3	5
6	6	15	21
7	7	4	9

Table 8.1 (continued)
Channel Number/Time Slot Assignments

Time Slot	Channel Number Assignments		
	D3/D4 Channel Bank	*D1D Channel Bank*	*D2 Channel Bank*
8	8	16	15
9	9	5	3
10	10	17	19
11	11	6	7
12	12	18	23
13	13	7	11
14	14	19	14
15	15	8	2
16	16	20	18
17	17	9	6
18	18	21	22
19	19	10	10
20	20	22	16
21	21	11	4
22	22	23	20
23	23	12	8
24	24	24	24

Besides DS1 SF compatibility, the DCS 1/0 has the capability to cross-connect ESF and clear channel (B8ZS) signals.

8.7.2 Cross Connections

Typically, the DCS 1/0 cross-connects DS0, bundled DS0, multiple DS0, and DS1 in any of the following five ways:

- *Two-way connections* provide bidirectional connection of an incoming channel to any vacant outgoing channel.

- *Conference bridging* provides the bidirectional connection of multiple channels.
- *Broadcast connections* provide the connection of an incoming channel to multiple outgoing channels.
- *One-way connections* provide the connection of an incoming channel to any vacant outgoing channel.
- *Loopback connections* provide the bidirectional connection of a channel back to itself.

8.7.3 User Interface

Access to the DCS 1/0 is either through a local terminal or a remote terminal linked to the frame via a modem or an X.25 communications connection. A common language used to send commands to the DCS 1/0 is the man-machine language (MML). Though different machines may use MML, there are always slight differences in the implementation from vendor to vendor. With the popularity of the newer transaction language 1 (TL-1), some vendors may be developing a TL-1 interface for their machines as well.

There are various ways to classify commands. One set of commands is sent by the user and the digital cross-connect responds. Commands to configure the system, to perform cross-connects, to perform testing and pre-provisioning, etc., fall under this category. Another set of commands is under the control of and is originated by the machine itself, either periodically or in response to some internally generated message. Examples are automatic commands to the diagnostic subsystems either on a periodic basis or in response to internal fault messages, status information, or alarms. Still another set of commands is comprised of the software commands controlling the functions of the computer central processors. More discussion on network operations can be found in Chapter 11.

8.7.4 Software Operating Systems

The DCS 1/0 software runs under the control of an operating system. The operating system performs many functions, among which are prioritizing and scheduling different essential tasks. These include managing the application software and the various system memories, including the hard disk drives often used to store the database and the performance data.

8.7.5 Alarm Interface

Alarm information is normally provided to the local DCS 1/0 site and to the remote administration center surveillance equipment. System alarm levels are usually classified as critical, major, and minor. A fourth alarm level is processor alarm.

A critical alarm occurs when slip or frame losses reach the preset service affecting threshold level on 5 or more DS1s or when there is no active switching matrix.

A major alarm can occur when one clock reference is lost or when the number of slips or frame losses of one digroup reaches the preset service affecting threshold level.

A minor alarm can occur when a redundant module fails or when the number of slips or frame losses of one digroup reaches the preset maintenance threshold level.

A processor alarm occurs when one or more critical system administration central processors fail.

Alarms are usually manifested in three ways:

- Alarms are indicated through LEDs mounted on front panels located in the DCS frame itself.
- Alarms relay contact closures, which in turn activate other equipment, either onsite or at remote sites, to inform network personnel of any problems with the system.
- Alarms are conveyed via electronic messages informing local or remote personnel of the system failures. These messages usually appear on terminals either as text or as a graphical presentation if further data processing has been performed.

8.7.6 Maintenance

Maintenance refers to two classes of activities concerned with performance monitoring and upkeep. One ensures maintenance support for the digital cross-connect. The other ensures the health of the facilities terminating on the DCS frame.

Besides the alarm features discussed in the foregoing section, the administrative subsystem continuously checks on the health of the frame as part of its self-diagnostic feature. It can provide fault detection, diagnostics, reconfiguration, software for error recovery, and audits to help section off the problem within the system.

The facility maintenance functions comprise facility alarms and performance monitoring, including slips, loss of frame, and performance thresholds.

Once problems are known, operations personnel should be able to initiate more detailed diagnostics on equipment or facilities that have failed or are providing substandard performances.

8.7.7 Test Access

Any DS0 terminating on the digital cross-connect must be able to be connected to the test access paths (TAP) located on a designated test access digroup

(TAD) termination to monitor and test the circuit. Under the monitoring mode, the cross-connect must be left intact and only bidirectional bridging of the test circuit to the TAP can be performed. Tear down of the test circuit with the split or terminate and leave command is sanctioned only after monitor access had been completed. Testing can then proceed with the insertion and measurement of test signals.

Release of the circuit to the original cross connection from either the monitor or split configuration must occur with a single command.

Test access to a DS1 signal is through a similar facility called the facility access digroup (FAD). The FAD features are similar to the TAP features, but FAD operates at the higher DS1 bit rate and is composed of two DS1 ports.

8.7.8 System Security

The DCS often maintains security with features that prevent unauthorized users from logging into the machine. Even after logging in, the user can be restricted from command-level access.

To access the DCS 1/0 system, a user must successfully complete a logon sequence that consists of entering a user name and password. For some systems, an additional callback sequence is provided to enhance the security, as described in Chapter 11. In addition to the logon and callback features that protect system access, the DCS 1/0 often has the capability to maintain command-level protection. This is provided by restricting users to a subset of the available commands, thus preventing unauthorized users from altering system operation.

8.8 APPLICATIONS OF THE DCS 1/0

8.8.1 General

There are many applications of the DCS 1/0, and its ultimate usage depends on the needs of the user of this versatile network element. Some are outlined here, and most of those listed are generic. However, a few of the uses may depend on special features that only some specific DCS 1/0 may have.

8.8.2 Back-to-Back PCM Channel Bank Replacement

The traditional way to groom, or to drop and insert DS0 channels from a DS1 signal stream, has been using channel banks back to back with the attendant demultiplexing, decoding to VF, routing through the test jacks, and finally terminating at the distribution frame. After hand rewiring at the distribution frame, the process is reversed until the DSOs in question are multiplexed into the DS1 signal again. A more efficient, faster, and practically error-free method is to use a DCS. This is illustrated in Figure 8.2.

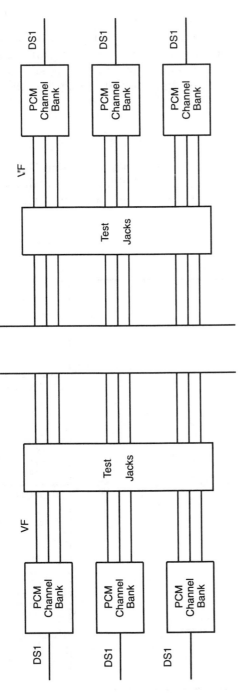

VF Frame

Conventional DS0 Cross-Connection Method

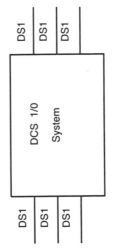

DS0 Cross Connection Using DCS 1/0

Figure 8.2 Back-to-back PCM channel-bank replacement.

In addition, the analog-to-digital conversion, which incurs some signal degradation, is avoided when the DCS is used. Other advantages include having the connection and test access functions available electronically, which makes eliminating errors due to manual rearrangements and rewiring possible. Also since test access is available in the DCS 1/0, the possibility exist to eliminate the circuit test jacks at the distribution frame.

8.8.3 Circuit Grooming

The DCS 1/0 can be used to cross-connect DS0s from partially filled DS1 facilities to concentrate them on fewer outgoing transmission facilities, as illustrated in Figure 8.3. This is especially cost-effective if the outgoing facilities are expensive long-haul circuits.

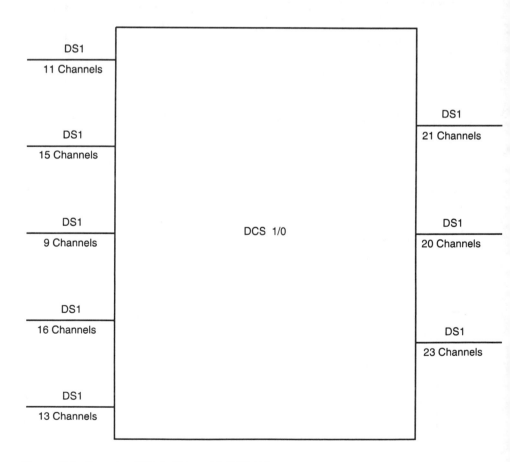

Figure 8.3 Grooming DS1 facilities with DCS 1/0.

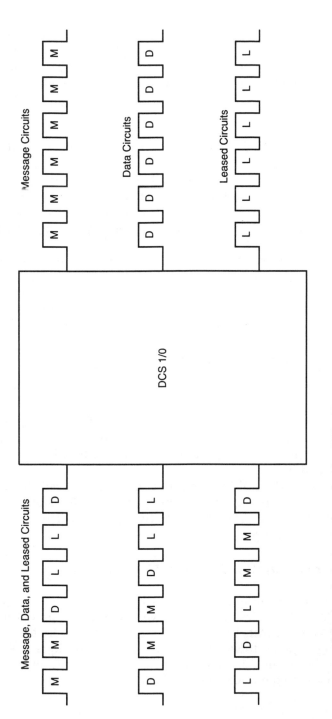

Figure 8.4 Grooming circuits by service type using DCS 1/0.

use is to cross-connect the DS0 circuits by traffic types. An example is to segregate the message, the data, and the leased circuits in the outgoing facilities as illustrated in Figure 8.4. The data circuits may be going to a corporate data center. The reverse function can be performed at the other end.

8.8.4 Hubbing

An important service of the DCS 1/0 is to provide a hubbing function to centralize a network. As shown in Figure 8.5, the flat or nonhubbing network on the left is inefficient, requiring direct circuit connections from each central office to the others. Such circuits are characterized by low fill and a greater number of facilities. It leads to high maintenance cost due to the allocation of more equipment and manpower to perform the engineering, testing, and provisioning at each site.

In the hubbing case shown on the right side of Figure 8.5, the central offices are connected only to a central site equipped with a DCS 1/0. The number of facilities is greatly reduced with a resultant improvement in the fill factor. Circuit engineering, reconfiguration, testing, and administration is simplified. The need to station personnel at the central office sites is diminished.

8.8.5 Remote Management Capability

The administration of cross connections, test access, maintenance, diagnosis, and DCS reconfigurations can be interactively performed locally or from remote operations centers. Access to the DCS 1/0 is through the access interface in the form of serial input/output administration ports as illustrated in Figure 8.6. For remote access, the administration ports are usually reached through dial up or dedicated lines.

The physical interface of the communications ports are RS-232C-compatible. The transmission baud rate varies from 1,200 to 9,600 baud with a default value usually at 1,200 baud. The administrative links can also be configured as X.25 ports each with multiple virtual circuits.

8.8.6 Customer-Controlled Network Management

Customer-controlled network management allows the user to access, to control the routing of his circuits, and to have surveillance of his portion of the DCS 1/0 network. This service lets the customer access the network through the network management system to control specific circuits. As illustrated in Figure 8.7, control may be through direct connection to the network management system from terminals or computers at the customer site.

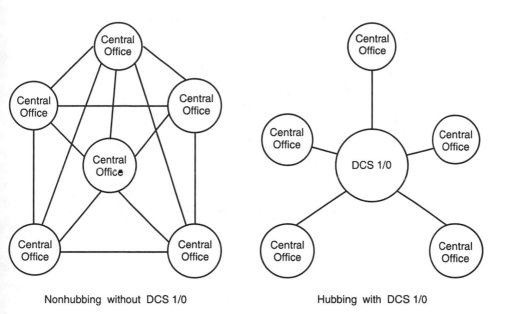

Nonhubbing without DCS 1/0   Hubbing with DCS 1/0

Figure 8.5 Nonhubbing and hubbing networks.

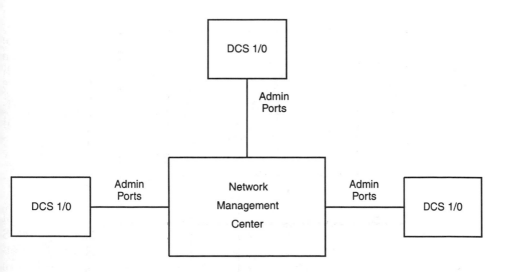

Figure 8.6 Remote management of DCS 1/0 systems.

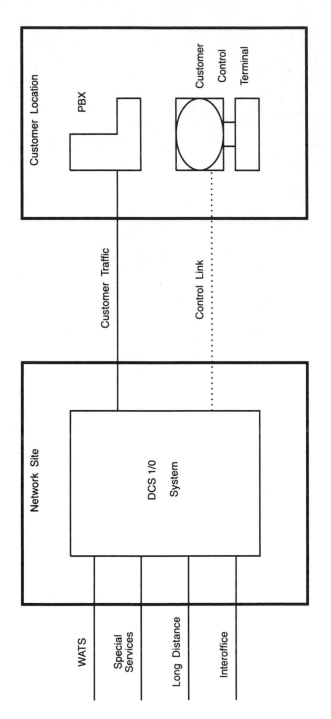

Figure 8.7 Customer-controlled network management.

8.8.7 Multichannel Switching

Multichannel switching allows the bundling of two or more DS0s as one entity. They are not separated as the group of DS0s transits through the network. Such a service is called fractional T1, and it allows the switching of multiple DS0s up to 24. The DCS 1/0 is well-equipped to handle this kind of service.

8.8.8 Digital Data System Hubbing

A digital data system (DDS) hub may be implemented using the DDS feature of the DCS 1/0 [2]. A separate set of optional feature cards provides the DDS functionality and allows access, test, cross-connects, and maintenance of subrate data circuits. The DDS features supported usually include the functionality of the sub-rate data multiplexer (SRDM), the multipoint junction unit (MJU), and the T1 data multiplexer (T1DM).

The conventional DDS hub makes use of multiple dedicated equipment types to implement it, as illustrated in Figure 8.8. Using the DDS feature of the DCS 1/0 greatly simplifies the implementation of an integrated hub, as shown in Figure 8.9.

8.8.9 Multipoint Bridging Services

Two very useful functions of the DCS 1/0 are the data-bridging and the conference-bridging features.

The split data bridging function provides bidirectional cross-connects between one master channel and multiple slave channels. The master can send to and receive from each slave channel. This is useful in point-of-sales or bank networks.

The conference bridging function provides bidirectional cross-connects between channels where all channels are treated equally. There is no master-slave relationship, as there is in the data conference bridge.

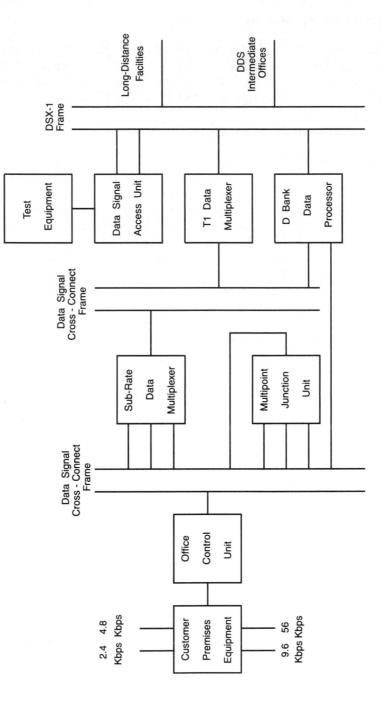

Figure 8.8 DDS hub using dedicated equipment.

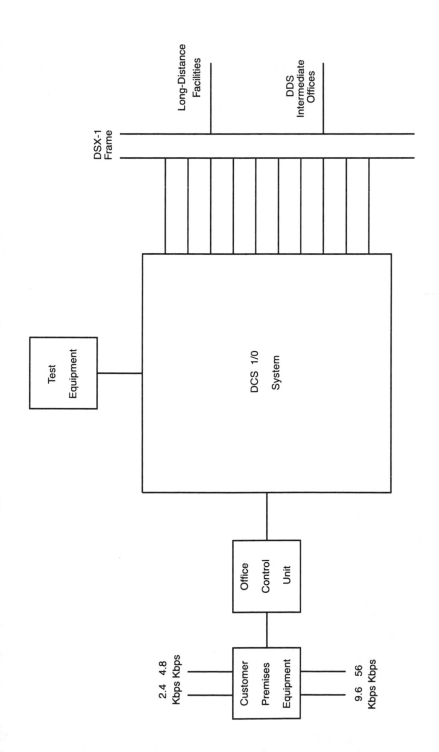

Figure 8.9 DDS hub replacement using DCS 1/0 system.

References

[1] Bellcore, Technical Reference TR-TSY- 000170, Digital Cross-Connect System Requirements and Objectives, Issue 1, Nov. 1985, and TA-NWT-000170, Issue 1, Dec. 1991.

[2] Bellcore, Technical Advisory TA-TSY-000280, Digital Cross-Connect Systems (DCS) Requirements and Objectives for Sub-rate Data Cross-Connect (SRDC) Feature, Issue 1, May 1985.

DCS 3/1 System

9

9.1 INTRODUCTION

The DCS described in Chapter 8 terminates DS1 signals and rearranges DS0 signals, leading to its designation as the DCS 1/0. The DCS covered in this chapter terminates DS3 signals and rearranges DS1 signals and is designated the DCS 3/1 [1]. These centrally controlled network elements usually include the additional capability of terminating DS1 facilities and may also allow DS3 signals to be passed intact between the input and output ports. The DCS can electronically cross-connect, in a non-blocking manner, any DS1 in any DS3 facility to any available DS1 slot in another DS3.

The DCS 3/1 system plays a vital role in the evolving SONET [2,3] world, which is discussed in detail in Chapter 13. DCS 3/1 systems equipped with optical interfaces are being developed, and these will interconnect directly with SONET fiber-optic systems at the optical-line level. To ensure SONET compatibility, the switching matrix of a DCS 3/1 system has to operate at a rate that will allow 1.728-Mbps SONET VT1.5 signals, as well as DS1 signals, to be cross-connected. This capability will be valuable in effecting a smooth transition from an asynchronous network to a SONET network. Any system that does not have this capability will have more limited applications in SONET networks.

9.2 DESCRIPTION OF THE DCS 3/1

DCS 3/1 systems can, like the DCS 1/0, be divided into four subsystems as shown in Figure 9.1, which depicts the overall block schematic. The subsystems are:

- Administrative subsystem;
- Input/output interfaces;

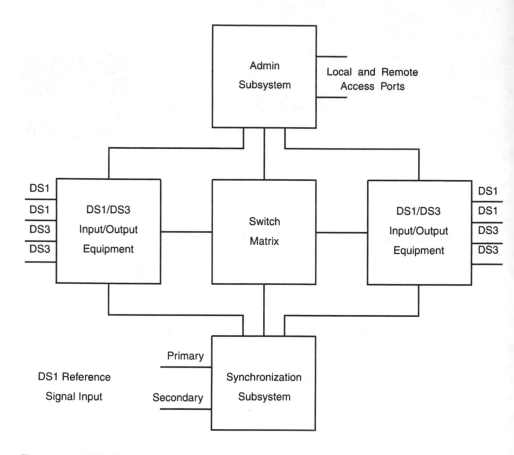

Figure 9.1 DCS 3/1 system block schematic.

- Switch-matrix subsystem;
- Synchronization and timing subsystem.

9.3 ADMINISTRATIVE SUBSYSTEM

The administrative subsystem provides the intelligence of the system and acts as the clearinghouse for all DCS 3/1 activities. The functions provided by the subsystem include the continuous status monitoring and diagnostics of all the circuit cards, the provision of user interfaces, the reception of user input information and commands, the initiation of the actions needed to respond to that information, the initiation of cross-connects, the management of the system database, and the initiation of protection switching and system-alarm indications.

The microprocessor-based administrative subsystem has both volatile and nonvolatile memories. The backup nonvolatile memory consists normally of EPROMs, hard disks, or streaming tape. Floppy drives are common methods used to load software into the DCS 3/1.

9.4 INPUT/OUTPUT INTERFACES

Both DS3 and DS1 interface ports are provided for the asynchronous traffic signals. Each DS3 facility and each block of DS1 facilities are multiplexed into an internal common format in a high bit-rate stream proprietary to the system vendor. These high bit-rate streams are applied to the switching matrix and at the far end are demultiplexed back into the output DS1 and DS3 facilities.

The input/output (I/O) equipment and the switch-matrix equipment are in many cases mounted on separate bays, and the cabling between them carries the high bit-rate stream. One of the ways to greatly reduce the cabling congestion and the resultant problems of crosstalk coupling inherent in this separation is by using fiber-optic cables at this interface. This allows the bays mounting the I/O ports to be up to 200 ft away from the switch-matrix equipment.

The I/O ports are protected against module failure on a 1:N basis, commonly with $N = 31$. In some systems, the switch matrix may support up to 256 DS3 ports, but only some of these may be available for facility connections, with the remaining ports assigned to protection.

9.5 SWITCH MATRIX SUBSYSTEM

The DCS 3/1 system switching matrix may be implemented with multiple stages of space switching, time switching stages using time slot interchange (TSI), or a combination of space and time switching. The high bit-rate stream from the I/O interface modules is applied to the matrix at a rate that will allow SONET signals to be cross-connected in future SONET applications. The minimum bit rate that can be chosen is the 1.728 Mbps of the VT1.5 signal. The matrix configuration allows the DCS 3/1 system to provide nonblocking connections for any DS1 signal in any incoming DS3 stream or any DS1 input facility, to any DS1 allocation in any outgoing DS3 stream or to an outgoing DS1 facility.

The maximum port capacity of a DCS 3/1 system is limited by the size of the matrix. Some systems have an initial configuration with a maximum capacity of 256 DS3 ports or a mixture of DS1 and DS3 ports with an equivalent of no more than 256 DS3 ports. These systems may be expanded to a larger port-capacity ports, as discussed in Section 9.7.

9.6 SYNCHRONIZATION AND TIMING SUBSYSTEM

For asynchronous applications of the DCS 3/1, no external synchronization input signals are required, and timing is derived from duplicated internal clock signals.

Synchronous operation, which will be necessary when SONET systems are connected to the DCS 3/1, requires the use of external timing references. The system clocks are locked to the network precision-reference frequency by means of duplicate DS1 signals, which act as primary and secondary sources.

9.7 DCS 3/1 EXPANSION

9.7.1 Growth to 256 Ports

The basic versions of DCS 3/1 systems have a maximum capacity of 256 DS3 ports, or the equivalent in DS1 and DS3 ports. As initially installed, the systems are wired and equipped to meet the current traffic demands, with an allowance for short-term growth in facility requirements. The first stage in expansion is the addition of DS1 and DS3 port modules, and in some cases switch-matrix modules, until the bays provided with the initial installation are filled. If necessary, additional bays providing port-module capacity may be added until the maximum capacity of 256 DS3 ports is reached. The growth up to 256 ports is achieved with no disturbance to the traffic already being carried on the system, and this is called in-service capacity expansion.

9.7.2 Growth beyond 256 Ports

In sites where the traffic demand exceeds the maximum capacity of 256 ports and a larger capacity DCS 3/1 system is not yet available, multiple systems can be installed to increase the total port count. In order to retain complete non-blocking cross-connect capability, some of the ports on each of the multiple systems must be used as interties between systems, and this reduces the total quantity of ports available for connection to traffic.

When equipment supporting more than 256 ports becomes available, the expansion of a system requires the replacement of the original switch matrix with a larger version and the addition of I/O interface equipment. In addition, the original administrative subsystem equipment may need to be replaced by a version capable of handling the larger volume of transactions inherent in the larger system.

The detailed procedure for replacing the switch matrix on an in-service basis is different for each vendor's system, but the activities are similar. The automatic 1:1 protection switching of the original matrix is disabled, and the traffic is locked onto copy 1 of the matrix. Matrix copy 2 is either replaced or modified to provide the larger capacity. After testing, the existing cross connections are mapped into copy 2, and the traffic is switched from copy 1 to copy 2. Matrix copy 1 is then replaced or modified and tested in the standby condition. When testing is complete the automatic-protection switching function is restored.

By these means expansion beyond 256 ports is made on an in-service basis. The cross connections cannot be changed during the transition, and the system operates in an unprotected mode.

9.8 TRADITIONAL DCS 3/1 FUNCTIONS

This section describes some important traditional DCS 3/1 functional characteristics. There are fundamental characteristics shared by all DCS 3/1 systems, especially if they comply with the Bellcore recommendations. Nonetheless, variations exist from one equipment manufacturer to another. Many of these functions are similar to those of the DCS 1/0 described in Chapter 8, and only those that are appreciably different in the case of the DCS 3/1 are described here.

9.8.1 DS1 and DS3 Compatibility

As in the DCS 1/0 system, the DCS 3/1 internal processing is independent of the channel time slot assignments within the DS1. The DCS 3/1 also provides ESF and clear channel capabilities on DS1 lines. The DS3 ports of the DCS 3/1 can accommodate the M23, SYNTRAN, and C-bit parity formats described in Chapter 6.

9.8.2 Cross Connections

The DCS 3/1 cross-connects DS1 to DS1, DS1 to DS3, and in some cases DS3 to DS3 nonchannelized signals, in any of the following four ways:

- One-way connections – an incoming channel is connected to any outgoing channel;
- Two-way connections – an incoming and outgoing channel is connected to any other incoming and outgoing channel;
- Broadcast connections – an incoming channel is connected to multiple outgoing channels;
- Loopback connections – a channel is connected back to itself.

9.8.3 User Interface

User access to the DCS 3/1 is provided by using a set of command messages to configure the system, access DS1 and DS3 performance monitoring data, and display information about the system.

9.8.4 Test Access

Test access is provided to both DS1 and DS3 signals. A complete range of test commands can be used to facilitate monitoring, split access, and terminate and leave test configurations through a local or remote terminal. Test signals can then be inserted and measured.

9.9 APPLICATIONS OF THE DCS 3/1

9.9.1 General

There are a number of possible applications of the DCS 3/1. Its ultimate usage depends on the needs of the user of this versatile network element. Some are outlined here and most of those listed are generic, but a few of the uses may depend on special features that only some specific DCS 3/1 may have.

9.9.2 Back-to-Back M1-3 Replacement

For drop and insert of DS1 facilities from DS3 signals, the DCS 3/1 is more efficient than conventional back-to-back M1-3 methods. Figure 9.2. illustrates the two cross-connect methods; the first using back-to-back M1-3 multiplexers, and the second using a DCS 3/1. Using the back-to-back M1-3 cross-connect method, the entire DS3 must be demultiplexed and decoded into 28 individual DS1 signals before any facilities can be dropped or inserted. The DS1 signals terminate on the DSX-1 distribution frame, where hard-wired cross connections are performed manually. Each facility reconfiguration, reconnect, or disconnect is labor-intensive. After the connection has been made at the DSX-1 frame, the DS1 signals pass to the second M1-3, which modulates the 28 DS1 signals into the 44.736-Mbps DS3 outgoing line signal.

The DCS 3/1 is advantageous for drop and insert procedures because all cross-connection and test-access functions are accomplished through software commands—thus eliminating the DSX-1 frame and the manual process of hard wiring the connections. The elimination of the DSX-1 frame allows the distance between the DS1 ports of the DCS 3/1 system and DS1 ports on other network elements to be doubled to 1,310 ft.

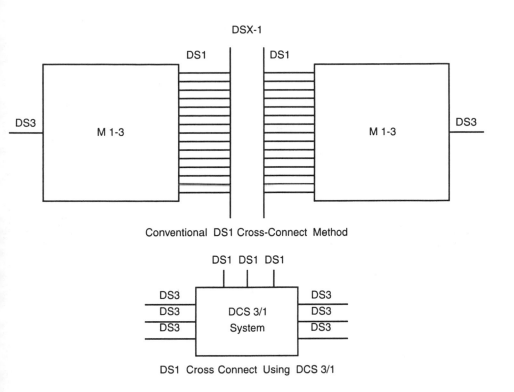

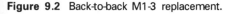

Figure 9.2 Back-to-back M1-3 replacement.

9.9.3 Facility Grooming

The DCS 3/1 may be used to groom DS3 facilities, as illustrated in Figure 9.3. The DCS 3/1 can cross-connect signals from a number of incoming partially filled DS3 facilities, and it can consolidate these signals onto fewer outgoing facilities. This application improves the utilization of costly long-haul facilities.

9.9.4 Access Traffic Grooming

Traffic gains access to a long-distance network at a point of presence (POP) and is carried on digital facilities that have replaced the voice-frequency signals used with analog networks. The digital facilities are at the DS1 or DS3 level and are carried on metallic cables, although fiber-optic cable systems may be used where a large volume of traffic is involved. As the SONET technique is introduced into networks, the access traffic will be provided over SONET links.

No matter which form of access system is used, the traffic on it is multiplexed into facilities that are convenient and economical to the local telephone company providing the traffic. Access trunk groups are provided based on the

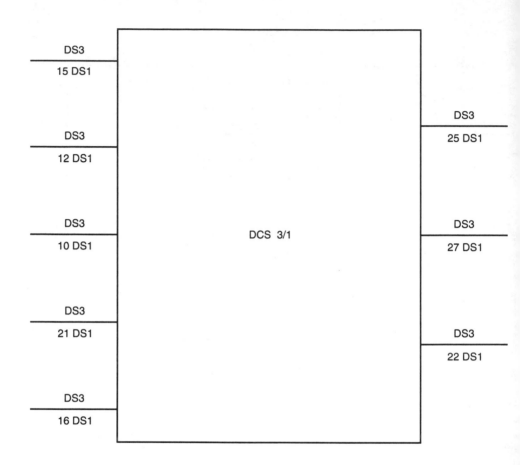

Figure 9.3 Grooming DS3 facilities with DCS 3/1.

local switching plan and on the local network routing to end central offices. Nonswitched service traffic is multiplexed to use the minimum number of facilities in the telephone company's plant. The access traffic that enters the long-distance network requires grooming into facility plans that are convenient and economical to the long-distance system operator.

There are two basic approaches that can be taken to groom access traffic. In the first of these, the traffic is not groomed at all at the access site but is carried over a transmission path to a major site for the necessary sorting into traffic types and rerouting towards its destination. In the second approach, the traffic is groomed at the access site into facilities that require a minimum of grooming at major sites. Because transmission systems accept and deliver traffic at the DS3 level, the grooming process involves rearranging DS1 signals so that they are placed in the optimum DS3 facility. A DCS 3/1 system is ideal for this purpose.

The first approach is illustrated in Figure 9.4. Access facilities at the DS3 level are connected to access-line systems and transported to a major junction site. A large-capacity DCS 3/1 system is used to perform the necessary grooming, which, prior to the introduction of DCS 3/1 systems, would have required very many M1-3 multiplexers and a large DSX-1 frame. This method conserves equipment, space, and power at the access sites and uses a minimum amount of access-line system capacity. The limiting factor for this approach is the maximum port capacity of the DCS 3/1 system at the junction site.

Figure 9.5 shows the use of relatively small capacity access DCS 3/1 systems for grooming at the access sites, prior to connecting the DS3 facilities to the line systems. Because the bulk of the grooming is performed at the access site, the DCS 3/1 system at the major site requires fewer ports than in the first approach, and the available port capacity is less likely to be a limitation on network design. As SONET interfaces become available for the DCS 3/1, the configuration in Figure 9.6. may be used. The line system terminals disappear because the DCS 3/1 provides the line terminations, and SONET systems are used for the access traffic from the local telephone company. The size and power requirements of the equipment at the access site diminish greatly, and remote control and monitoring at these sites is simplified.

9.9.5 Fractional DS3 Services

Fractional DS3 services occupy a number of adjacent DS1 allocations in a DS3 facility. Multiple DS1 cross connections may be made in a DCS 3/1 system, and by cross-connecting the correct DS1 signals in an incoming DS3 to the correct DS1 slots in an outgoing DS3 facility, a fractional DS3 service may be handled by the DCS 3/1 system.

9.9.6 International Traffic Transport and Routing

Long-distance networks in the U.S. adhere to North American standards and are designed to carry traffic originating and terminating within the region. They are, however, called upon to carry traffic that gains access to the networks at international gateway points and is transported from one gateway to the other with no access at any intermediate locations. International links that appear at gateway points include trans-Atlantic and trans-Pacific cable systems, and trans-border route crossings from Mexico.

Digital traffic that originates in countries that follow the CEPT standards uses the 2.048-Mbps E1 format for the first multiplexing level instead of the North American 1.544-Mbps DS1 stream. These E1 facilities can be carried over a US asynchronous digital network by multiplexing three E1 signals each at 2.048 Mbps into one 6.312-Mbps DS2 stream in a modified M1-3-type multiplexer (called M1-3') with E1 interfaces. The DCS 3/1 systems do not make any

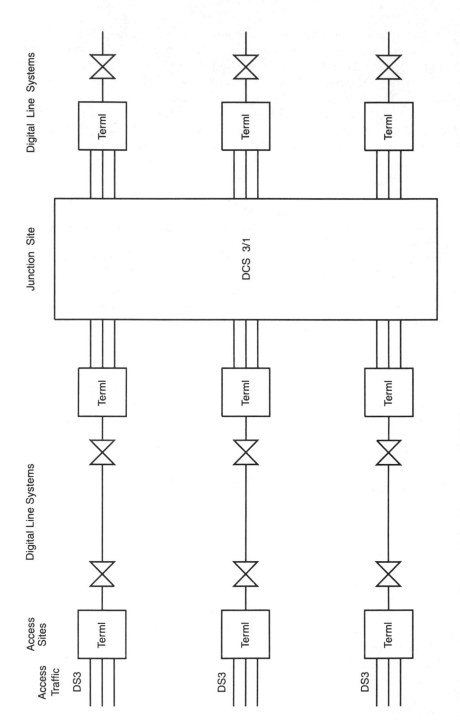

Figure 9.4 Access traffic grooming without access DCS 3/1 systems.

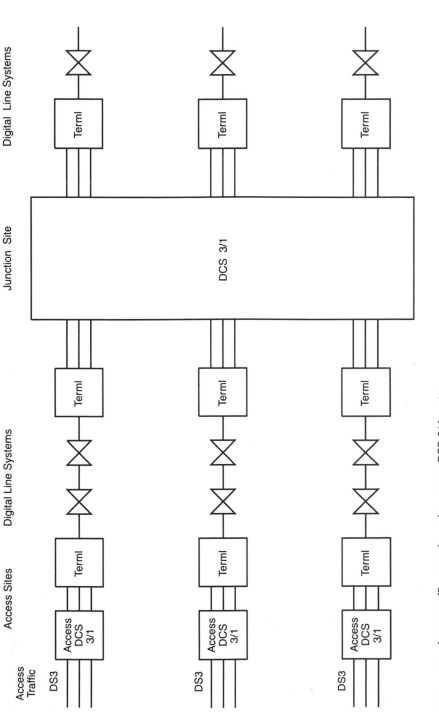

Figure 9.5 Access traffic grooming using access DCS 3/1 systems.

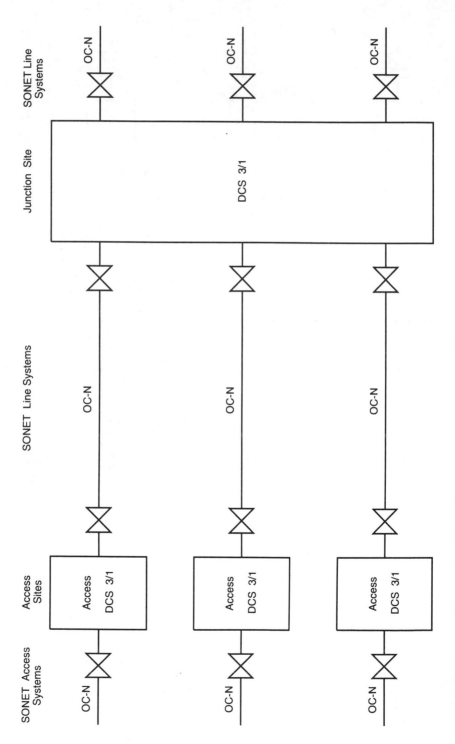

Figure 9.6 Access traffic grooming using DCS 3/1 systems with SONET interfaces.

provisions for the direct cross connection of DS2 signals but can carry intact the DS3 containing E1s.

In a SONET-based network, one E1 signal can be mapped into a VT2 virtual-tributary stream, or three E1 signals can be placed in a VT6 stream. These VT signals can then be mapped into part of the payload capacity of an STS-1, which can be transported on the SONET network.

The STS-1 signals that carry the E1 facilities may need to be routed through the network from gateway to gateway. A DCS 3/1 system equipped with SONET interfaces and with the ability to switch an STS-1 signal intact from input port to output port can be used to route these facilities, without any access to the E1 traffic along the route.

References

[1]   Bellcore, Technical Reference TR-TSY- 000233, Wideband and Broadband Digital Cross-Connect Systems Generic Requirements and Objectives.

[2]   Bellcore, Technical Reference TR-TSY- 000253, Synchronous Optical Network (SONET) Transport Systems: Common Generic Criteria.

[3]   ANSI T1.105-1991, American National Standards for Telecommunications, Optical Interface Rates, and Formats Specifications (SONET).

DCS 3/3 System

10

10.1 APPLICATIONS

The DCS 3/3 system is a remotely controlled DS3 cross-connection device [1,2]. One of its major applications is as a replacement for the manual DSX-3 frames currently used in the connection, disconnection, and rearrangement of DS3 facilities in an office. Many end-to-end DS3 facilities are made up of segments interconnected via jumper cables on the DSX-3 frames at a number of sites. Preparing a facility for service involves activities at each intermediate site to cable and test the DS3 path for continuity. The DS3 facility implementation has to be scheduled for a time when personnel are available at all of the sites simultaneously, and this usually means a considerable delay between the completion of the facility planning phase and the placing of the facility in service.

The DCS 3/3 system provides the ability to add or remove any DS3 cross connection by remote-control command from a central site. This avoids the need for human intervention at any of the intermediate locations involved in a DS3 facility implementation. Personnel are needed only for facility testing at the end sites of the DS3 path. The time involved in placing new or rearranged facilities in service by this method is minimal, and the problems encountered when mistakes are made using manual cross connections on the DSX-3 frame are avoided. In addition to the savings in labor costs gained by using the DCS 3/3, the ability to react to urgent demands for addition or rearrangement of DS3 network facilities is a great advantage. The reliability of the network facilities is also improved by the deletion of the DSX-3 frames and the associated coaxial connections.

Another application of the DCS 3/3, which makes use of its remotely controlled reconnection capabilities, is in the survivability of DS3 access traffic, as described in Chapter 12.

10.2 GENERAL DESCRIPTION

Unlike the DCS 1/0 and the DCS 3/1 systems described in previous chapters, the DCS 3/3 system performs no multiplexing functions, and it is therefore a less-complex device. As shown in the block schematic in Figure 10.1, it has only DS3 ports, any two of which may be cross-connected through the matrix on a one-way or two-way basis. Broadcast connections are also possible, with one input port connected to more than one output port. The DCS 3/3 system is transparent to the DS3 bit streams passing through it. It does not require external synchronization because each cross-connection path derives its timing from the DS3 input signal. DCS 3/3 systems are available with maximum capacities ranging from 60 to 1,920 duplex ports, so a variety of site configurations may be implemented. Any of these systems may be equipped initially with less than the maximum number of ports and expanded by adding ports as the DS3 traffic increases.

The system consists of three major parts: the DS3 port equipment, the matrix switch, and the administrative subsystem. An optional feature is performance monitoring of the DS3 traffic. The remainder of this chapter provides details of the functions and features of DCS 3/3 systems, including equipment protection arrangements.

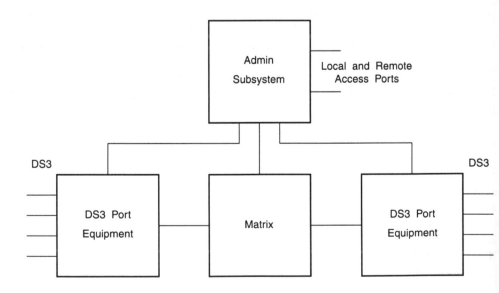

Figure 10.1 DCS 3/3 system block schematic.

10.3 DS3 PORT EQUIPMENT

Each port module includes both input and output circuitry, and the DS3 interface has to meet the parameters given in the standards [3]. A line build out (LBO) device is provided at the DS3 output point to allow for installations where the cable distance to the adjacent network element is less than 900 ft of Western Electric coaxial cable type 728A or equivalent.

At the DS3 input point, an automatic line build out (ALBO) equalizes the received pulses and is followed by a level detector, which decides whether or not a particular bit time is occupied by a pulse. The timing of the sampling instant is derived from the incoming pulse stream, and the result is a regenerated *clean* DS3 rate signal. The bit error rate (BER) measured from DS3 in to DS3 out does not exceed 1×10^{-10} as specified by Bellcore [2]. Measured BER values of better than 1×10^{-12} are normal, but care must be taken in the installation of the system to avoid unwanted coupling due to grounding problems and crosstalk between the cables.

The reverse path takes the DS3 bit stream from the matrix, adjusts the level of the pulses, and passes them via the LBO to the output port. An activity detector is provided to sense the voltage level of the pulse stream. It is arranged to cause an alarm if the signal falls below a minimum level.

10.4 MATRIX

The matrix consists of three stages of switching, as shown in Figure 10.2. The input stage has N input ports and $2N$ output connections to the center stage. The output stage has 2N input connections and N output ports, and the complete matrix provides nonblocking connections between any input port and any output port. When fully equipped with switching elements, the matrix can support the maximum number of DS3 ports on the system, but in some systems the matrix may be expanded in modular steps to match the growth in equipped DS3 port capacity.

The matrix operates at a speed greater than the DS3 rate of 44.736 Mbps to allow the addition of overhead bits to the DS3 bit stream. These added bits are used to verify that the correct ports have been joined together and to allow

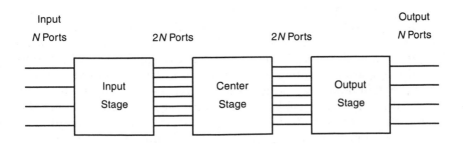

Figure 10.2 DCS 3/3 matrix block schematic.

internal testing of the matrix. In order to prepare for the deployment of the SONET systems [4] described in Chapter 13, the matrix rate of DCS 3/3 systems deployed in networks actually exceeds the SONET STS-1 rate of 51.84 Mbps.

10.5 ADMINISTRATIVE SUBSYSTEM

This function of the DCS 3/3 contains the operating-system information and the database relating to system provisioning, equipment status, and the current DS3 cross connections. The system memory may be distributed between the DS3 interface equipment and the administrative modules. The memory elements associated with the DS3 interfaces contain the existing cross-connect status to aid in the rapid implementation of changes to the cross-connect conditions. The system operating programs, together with the current database, are stored in nonvolatile memory elements in the administrative subsystem so they will be retained in the event of a loss of power to the DCS 3/3.

The administrative subsystem includes the access ports to the local and remote-control terminals. Software programs are loaded into the system when it is being prepared for service, and these programs provide the operating procedures for the cross-connect. Commands are entered using local or remote-control terminals to add, delete, or modify connections between the ports and to retrieve information about the system state. Alarm and status messages are generated spontaneously and are displayed on the control terminal screens, as well as at the remote control and status center. Access to the system commands is restricted to authorized operators through the use of individual user-identification names and passwords.

10.6 PROTECTION

The purpose of the protection provided in a DCS 3/3 system is to avoid a loss of DS3 service caused by equipment failure. An alarm is given when an incoming DS3 facility signal is lost, but the DCS 3/3 cannot protect against loss of service when the failure is due to external causes.

The DS3 port equipment is protected on a 1:N basis, with one protection circuit for N working paths. The maximum value of N is usually specified as 16. The equipment is designed so that if any one DS3 circuit within a 1:N protection group fails, the protection circuit takes over until the faulty circuit is replaced with a spare port module. When the DS3 facility reverts to the normal path, the protection circuit is again available for the next failure. It follows that only one path can be protected at a time, and a second failure within the same protection group is not protected. A major operational requirement is that the removal of any one DS3 port plug-in module is not allowed to cause loss of service, so the working and protection paths of any DS3 circuit cannot be mounted on a single plug-in module.

Protection of the matrix is also provided, and the basis varies in different systems between 1:1 and 1:31. The failure of any plug-in matrix card results in the cross connections that it was carrying being transferred to the protection-matrix circuitry. Again, the removal of any one plug-in matrix card is not allowed to cause loss of service.

The administrative subsystem is not protected because any equipment failure here does not cause disconnection of established cross connections. There may, however, be loss of communication with the control center, and it may be impossible to add, delete, or change cross connections until the faulty equipment is replaced.

10.7 PERFORMANCE MONITORING

Any DS3 facility that traverses the network passes through a number of transmission system segments and various network elements that may include DCS 3/3 systems. The purpose of facility performance monitoring is to isolate the cause of any failure or degradation of performance to a particular portion of the route.

The minimum provision of performance monitoring on a DCS 3/3 system is a loss of signal (LOS) indication on each DS3 input port. If that port is cross-connected to an output port, the LOS condition causes an alarm indication signal (AIS) to be transmitted from the output port. The AIS is detected at the input port of each succeeding piece of equipment in the path and is used to suppress any further LOS indications. This allows the fault to be isolated to the segment where the signal was lost.

More sophisticated performance-monitoring circuitry may be built into the DCS 3/3, allowing indications of loss of frame, excessive BER, errored seconds, and severely errored seconds. The number of occurrences of each parameter may be counted in 15-minute, one-hour, or daily registers, and the results may be displayed on the control unit by retrieving the performance-monitoring data. These parameters may have selectable thresholds so that if an excessive quantity of errored seconds is counted in one hour, for example, a threshold crossing alarm is given.

Devising a method of detection of errors in the received DS3 signal presents a challenge. Errors in the framing bits are seen at the DCS 3/3, but there are so few framing bits in a DS3 bit stream that only very poor BERs can be observed by this method. As detailed in Chapter 6, the original DS3 format, known as M23, does not provide means for in-service performance monitoring. However a later DS3 signal format, called C-bit parity, does allow the occurrence of parity errors to be registered, and performance parameters such as errored seconds and severely errored seconds can be calculated. Since the M23 and C-bit parity formats are not end-to-end compatible, both ends of a DS3 facility must use the C-bit parity format for the performance-monitoring technique to be used.

10.8 CROSS-CONNECTION MAPS

A list of the cross connections required to exist in a DCS 3/3 at any given moment is called a cross-connection map. Provision may be made for the storage of at least 100 maps, and no map needs to have any knowledge of the existence of any other map stored in the DCS 3/3. A map swap, which changes the cross connections from one map to another, is executed with a single command from the control terminal. The purpose of these maps is described in Chapter 12.

References

[1] Bellcore, Bell Communications Research, Technical Reference TR-TSY- 000233, Wideband and Broadband Digital Cross-Connect Systems Generic Requirements and Objectives.
[2] Bellcore, Bell Communications Research, Technical Advisory TA-TSY-000241, Electronic Digital Cross-Connect (EDSX) Frame Requirements and Objectives.
[3] ANSI T1.102-1987, American National Standard for Telecommunications, Digital Hierarchy-Electrical Interfaces.
[4] ANSI T1.105-1991, American National Standard for Telecommunications, Optical Interface Rates, and Formats Specifications (SONET).

Network Operations 11

11.1 INTRODUCTION

A major task of the network operations function is to maintain the integrity and performance of the network so that it can carry the revenue-bearing traffic assigned to it. Surveillance of the network status and alarms allows the identification of fault conditions and performance degradation so that corrective action can be taken to restore network integrity. Provisioning of equipment and facilities is implemented to meet new and changed demands for service, and testing allows the performance of new and repaired equipment to be verified.

All of these activities can be performed at the equipment sites, and the results can be collected and passed to the network management for action. However, it is more efficient to leave the equipment sites unstaffed, except when equipment reconfiguration or replacement is required, and to conduct the surveillance, provisioning, and testing remotely from one or more network control centers [1]. Many digital systems and network elements provide access, both local and remote, for software-controlled operation. Unauthorized access to the network is potentially disastrous, and various security precautions are taken to try to prevent this from happening.

11.2 NETWORK SURVEILLANCE

The simplest form of surveillance is the observation of the alarm conditions signaled on the network. For many years the alarms were displayed only at the individual equipment sites, and the site personnel had to identify the fault condition causing the alarm indications. Complicated problems required the cooperation of people at more than one site to isolate the causes. It was difficult to sort out cause from effect, and a single equipment or transmission system failure could cause multiple alarms to be displayed at several sites. Each such incident was recorded in a site log, and these records were available to the operations management for analysis.

This form of surveillance could not provide an overall view of the alarm status of the sites. A great improvement in network surveillance occurred when the alarm conditions at the sites were encoded onto data links, allowing the information to be collected, and if desired displayed, at network control centers. In addition to equipment and facility alarms, the information encoded included the site housekeeping alarms such as *Door Open*.

The advent of software-controlled network elements and systems has made possible not only improved surveillance, but also remote testing and provisioning. The operations support (OS) techniques used with the different software-controlled systems varies, and the OS for the DCSs and for the SONET transmission systems are described later in this chapter. There are, however, some topics which are common to the operation of all the systems, and these are covered next.

11.3 SECURITY

Unauthorized access to the operations of a network element or system is potentially disastrous. It is possible for such a user to disrupt traffic on the network, either accidentally or maliciously, or to obtain information about a network that would be valuable to a competitor. To avoid this problem, it is usual for each authorized user to be given a personal identification (ID) and a password, to be used when logging on to the network with a local terminal or when logging on remotely through the operations support system (OSS).

The personnel whose assignments require them to have access have differing levels of responsibilities. The person with the highest level of user privileges is the system administrator, who must be able to request all reports on status and configuration; issue commands to provision and test network equipment and facilities; and to add, delete, and change users' passwords. The most basic level of privilege allows a user to access certain reports but not to make any changes. In general, each network element and system makes provisions for multiple levels of privilege, and each of these, apart from the highest, limits the user to a subset of the available commands.

Another security measure is available when access to the system is obtained over the switched network. This involves using a *callback sequence* when logging on. The terminal device used to access the system is equipped with an auto-answer modem. When the user logs on using his ID and password, the system immediately logs off the user and dials a pre-specified telephone number. The user's auto-answer modem responds, and the system makes another request for identification. When the user again enters the correct ID and password the logon is complete. This feature provides added security because it limits access to a user with the correct identification, who is also using a par-

icular terminal with an auto-answer modem which has the correct prespeci-
fied telephone number.

11.4 LANGUAGES, SYNTAX, AND MESSAGES

11.4.1 Languages and Syntax

Communication between the network operator and a network element requires
the use of a language and syntax that is unambiguously understood by both.
While a human is able to deduce the meaning of a message which differs from
the standard format, a machine can respond only to a message that conforms
exactly to the format stored in its memory. A wrongly inserted, misplaced, or
missing comma, for example, may make a message totally incomprehensible to
a machine.

Two languages in common use at present are man-machine language
(MML), and the Bellcore standardized transaction language 1 (TL-1) [2]. The
MML was developed earlier than TL-1 and is used on some DCSs, and TL-1 is
used on the other digital cross-connects and on SONET systems. More sophisticated languages are in development and may be standardized and implemented in the future.

11.4.2 Messages

Messages are of three types: commands, responses, and spontaneous reports.
Spontaneous reports are those that are not requested by the operator and include changes in status, occurrence of alarms, and regularly scheduled reports.
The format of commands includes the ID of the network element addressed, a
message serial number, a command verb such as *connect*, and the object or ob-
jects involved in the action, e.g., *port 12* and *port 46*. The last symbol of the
command denotes that the command is to be implemented. For those commands that would not otherwise be carried out due to the status of the network
element, the command includes an abbreviation for *forced*.

Responses to command messages include the date and time, the ID of the
network element, the message serial number, and a statement that the action
commanded has been completed or denied. In the latter case, a code word
showing the reason for denial is usually included. Spontaneous messages include the date and time, the ID of the network element generating the report,
and the status change, alarm message, or the scheduled report information.

11.4.3 Vocabulary

The use of a common language and syntax does not necessarily mean that a
message will be understood by the machine or person who receives it. Unless

the words that are used and the meaning of those words are common to both parties, the message may still not be understood.

An example is a command sent to the terminal equipment of a SONET system to switch the traffic from the active to the standby optical-output module. One vendor may have named this module the *OTU*, standing for optical transmit unit, while a second vendor may have used the term *SOU*, for send optical unit. Both of these designations are equally valid, but a command sent to the second system using the designation OTU instead of SOU will be denied. The situation is even more complicated where the packaging of the circuit functions is handled differently by different vendors. For example, one vendor may mount both the transmit and receive sides of a piece of equipment in one module, while a second vendor uses separate transmit and receive modules for the same function. In this case no common designations are possible.

For these reasons, it is necessary for the network operator to use the information on commands and responses included in the vendor's manual for each software-controlled network element and system to ensure that correct messages are used.

11.4.4 User-Friendly Software

The system operator's task may be greatly simplified if a vendor provides a user-friendly software package with a system. Using this package, the operator is not required to enter commands using the exact syntax and vocabulary for the network element being addressed. Instead, the operator is presented on the terminal screen with a menu of command elements from which to select. By following the procedure and making the correct selections from the choices shown at each stage, the command message is compiled and sent to its destination. This is of particular value when the operator is handling a number of different types of network elements supplied by more than one vendor. Some operators, however, become so familiar with the exact formats of the commands and responses for a system that they prefer to enter the commands directly, rather than follow the more time-consuming and cumbersome procedures required by the user-friendly menus.

11.5 DCS SYSTEMS

11.5.1 General

The major application of the DCS 1/0 system is in the provisioning of private-line circuits to be used between customer locations. These circuits may be at the DS0 level, or they may involve multiple DS0 blocks, or they may be connected on a DS1 intact basis. DCS 1/0 systems may be deployed at a large number of sites in a network, and an individual private-line circuit may traverse a number of sites in its end-to-end path. There is typically a large volume of activity in the provisioning, changing, and disconnection of these circuits. The tracing of a misrouted circuit is a tedious and time-consuming task, and implementing the circuit routings correctly is vital to the private-line provisioning activities.

The introduction of the DCS 3/1 system has relieved the DCS 1/0 systems of the need to handle all the DS1-level private service, since this newer system is more cost-effective in handling the DS1 facilities. In addition to this application, the DCS 3/1 provides remotely controlled cross connections making facility additions, rearrangements, and disconnections possible without the intervention of staff at intermediate points along the path.

The DCS 3/3 has two major applications. The first of these is to provide remotely controlled cross connections at intermediate points in a DS3 facility on the network. The second is to store maps, which are lists of DS3 cross connections, and swap from one map to another in response to a single command. The map-swap feature is used in the network-survivability method using network loops. In both of these applications, reliable and accurate DS3 cross connection is vital.

11.5.2 Access and Control

Several communications ports are provided on each DCS system for local access and, with the use of modem-derived circuits, for remote access.

Where an X.25 packet-data network is used for connection of the DCS systems to a network operations control center, X.25 communications interface ports are used. A packet-data network consists of a number of processors and communications links to which user terminals are connected. Data is sent over this network in packets, which have a defined format and a maximum packet length fixed by the particular version of X.25 that is supported. There are two types of X.25 logical circuits: the permanent virtual circuit (PVC), and the switched virtual circuit (SVC).

The PVC uses a long-term connection, and the connection data for the circuit entered by the system operator is maintained in the packet switch. Any subsequent configuration changes are also entered by the system operator. The packet used on a PVC includes the source address but not the destination.

The SVC offers increased usage efficiency of the packet network. A circuit is set up in response to service requests originated by a service point on the packet network. The source and destination addresses are included in the packet, and the connection is maintained only for the duration of the call.

Each end terminal employs a standard interface protocol at its interface with the network, and this conforms to CCITT Recommendation X.25. Multiple X.25 interface ports are provided on the DCS systems.

11.5.3 Surveillance and Alarms

Audible and visible alarms due to failures in the DCS equipment are signaled at the DCS system site. In addition, these alarms may be signaled on data links to the control center. Typically, four levels of alarms are provided. Critical alarms denote the loss of multiple ports and important equipment modules; major alarms indicate the loss of a small number of ports; and minor alarms are given for nonservice-affecting faults, such as the loss of one of a pair of redundantly protected modules. A processor alarm indicates a failure of the administrative function, which means that configuration changes are no longer possible, but the existing cross connections remain in place.

The occurrence of an alarm causes a spontaneous alarm report to be generated and sent to the control center. In addition, a command may be sent to the DCS requesting a report on the alarm status of the DCS system.

11.5.4 Performance Monitoring

The simplest form of performance monitoring is the loss of signal (LOS) indication given when the DS1 or DS3 signal disappears at the input port of a cross-connected facility. The LOS is accompanied by the application of an alarm indication signal (AIS) to the output port of the cross connection. An LOS condition causes a spontaneous-report message to be sent to the control center.

Where provided, enhanced performance monitoring allows one or more parameters of the incoming facilities to be examined. The ESF format for DS1 signals allows in-service monitoring of facility performance using the cyclic redundancy check bits. In the case of a DS3 facility, the performance data may be based on parity bits, F-bits, and M-bits or on C-bit parity. These parameters are described in Chapter 6.

The sets of performance data are stored in registers, typically in 15-minute and hourly blocks for the preceding 24 hours, with the daily totals retained in storage for the preceding 7 days. The performance criteria derived from the bit monitoring typically include errored seconds, severely errored seconds, out-of-frame seconds, and BER. Threshold values for each parameter may be set by the user, and threshold-crossing alert messages are sent to the control center when these values are exceeded. The performance monitoring data may be retrieved by issuing the appropriate commands to the DCS system, and automatically generated reports may be scheduled as required.

11.5.5 Port Provisioning

Port provisioning in DCS systems relates to the status of the DS1 and DS3 ports. In some systems, when an additional port module is installed, its presence must be registered by sending a command to enter it into the software memory. Commands are used to place a port in a *ready-for-service* state or, when maintenance of the port module is necessary, in an *out-of-service* state.

11.5.6 Cross-Connect Commands

There are three types of cross-connect commands used to implement one-way, two-way, and broadcast cross connections. Successful completion of the activity is indicated by a response message showing cross connection complete. A response message that shows that the command action was denied also includes a reason for the denial. This may show that one or more of the ports is in a failed or out-of-service state, or is already in service as part of another cross connection.

11.5.7 Map Commands

This type of command is used with the DCS 3/3 system. These systems are capable of storing at least 100 cross-connection maps, each of which consists of a list of cross connections for the system. A single command is used to swap from one map to another, e.g., *swap to map 65*. Any cross connections that remain unchanged from one map to another remain intact, with no break in DS3 transmission. The map swap is denied if any of the cross connections listed in the new map cannot be completed for any reason.

The reversal of traffic around a loop to restore service when a transmission link failure occurs is described in Chapter 12. This involves map swapping in multiple DCS 3/3 systems around the loop. A restoration plan for the action to be taken when a particular network link fails includes the appropriate map swap to be invoked at each of the DCS 3/3 systems involved. The restoration plans are maintained at the control centers, and the plans and maps are updated to reflect changes in the network DS3 facilities.

11.5.8 Testing

Provision is made for the testing of the facilities that are cross-connected in the DCS systems, and ports that are not in use for cross connections are selected under software control.

Bridging test access allows monitoring of the incoming signal, without disturbing the cross connection. When split test access is used, the cross connection is broken; the incoming signal can be monitored, and a test signal can be sent into the output port. When the command is issued to revert from split to bridging access, the original cross connection is restored.

11.5.9 DCS System Configuration

The DCS systems are remotely operated by personnel at a central location, and one of the tasks they perform is to establish the configuration of each DCS 1/0, DCS 3/1, and DCS 3/3 as it is brought into service. Since a large number of systems may be deployed in a network, it is advantageous that all DCS systems of one type be configured in the same way for ease of operation. This means, for example, that similar port cards in each DCS should be reserved for circuit and facility testing purposes. The port cards to be used for testing should be allocated in more than one of the available equipment shelves to reduce the risk of loss of testing capability due to common equipment failure. Similar care should be exercised in selecting the port cards that terminate the DS1 signals used for synchronization purposes, where applicable. Similarly, common port cards should be designated for broadcast or bridging services if such services are to be provided.

Where test equipment is provided with the DCS system, the same types of instruments should be supplied at each system site. In those cases where critically important services are being handled, the test equipment may be provided in a redundant configuration.

With the remote-operation capability established, it now becomes possible to concentrate scarce technical talent at one place instead of having it scattered at different sites. With a number of network technicians located at the control center, a certain amount of specialization in a particular type of system is possible. It is also more feasible for these technicians to provide technical backup when network performance problems have to be resolved.

11.5.10 Computer Files

Because DCS systems incorporate a large amount of computer technology, the systems exhibit a high degree of *smartness*. Nonetheless, it quickly becomes apparent to anyone familiar with DCS systems that using them can be very tedious. Another aspect of this is the repetitive nature of the commands or strings of commands employed to perform the various functions. Examples of such functions are the configuration of multiple systems at a number of network sites, moving the DCS commands in each of the systems into different designated classes, and the repetitive performance of circuit or equipment testing, which requires sending commands to both test sets and the DCS systems.

Simple computer files, called script files, work well for partial automation of many tedious and repetitive tasks. Use of the same set of computer files assures uniformity of results in configuration acceptance or routine testing of equipment. Circuit cross-connects can also be verified for accuracy before actual implementation into a DCS system.

11.5.11 DCS System Database Backup

For safety reasons, there is a crucial need to backup the database in the hard disk drive in case of drive failure. A daily automatic backup of the database is a highly recommended practice. The hard disk backup process should also be invoked on an unscheduled on-demand basis whenever significant database changes have taken place.

11.6 STRESS TESTING OF DS1 FACILITIES

11.6.1 General

The evolution of the DS1 signal from the original T1 system format to the superframe (SF) and to the extended superframe (ESF), is traced in Chapter 5. The original T1 system had no provision for in-service performance monitoring, and intrusive out-of-service testing was used to identify system problems. Testing of T1 systems was performed only when they were being placed in service, when they had failed, and for verification after completion of repair. Both ends of a system were equipped with PCM channel banks, and speech traffic predominated on the channels.

A long-distance DS1 path from customer to customer now typically traverses local-area systems and a number of interconnected DS1 facilities in the long-distance network. The signals may pass through T1 regenerators, M1-3 multiplexers, DCS systems, and a variety of digital transmission systems. The signals may carry a number of communications services in addition to speech traffic. The intensely competitive market for these services has generated customer demand for reliable error-free performance and changed the requirements for testing from a reactive to a proactive stance. This is exemplified by a move from the passive collection of data and correction of problems after failures have occurred to the monitoring of facility performance and the stress testing of a facility before it is placed in service. The objective is to identify network elements with marginal performance and perform corrective maintenance before network problems occur.

Comprehensive testing of DS1 facilities consists of injecting predetermined digital signals, called stress patterns, to test the performance-capability margins of the path. Various bit patterns are used to stress different circuit parameters, as explained in the discussions that follow.

11.6.2 Performance Monitoring and Testing

The complete failure of a DS1 facility results in network alarm indications and requires rapid corrective measures by the carrier to restore service. Deterioration of service quality may be detected by the customer as poor-quality speech or errors in data transmission. Both complete and partial service failures are a form of performance monitoring but are perceived as a lack of reliability or service, which may result in the customer moving to a different network supplier. The ESF format allows in-service performance monitoring of DS1 facilities. A deterioration in quality may be detected by the network operator and corrective action may be taken before it becomes apparent to the customer.

The deterioration of facility performance may result from problems in any one or more of the various network elements traversed by the facility. Testing a facility from end to end may reveal the existence of performance degradation but it is then necessary to test the segments of the facility individually to locate the source of the problem. The performance of a particular facility segment may vary according to the bit pattern of the DS1 signal being carried, and some network elements may show problems with one type of bit pattern that are not apparent when it is tested with a different signal. This means that there is no universal test signal that can be used to identify every type of problem in all DS1 facilities, and for this reason the various stress patterns and their applications have been devised. Rather than waiting for potential problems to appear, stress testing is being undertaken by network operators as part of the regular maintenance program, as well as prior to placing a facility in service.

11.6.3 Stress-Testing Patterns

Various testing patterns are shown in Table 11.1 and are used to identify specific types of system malfunctions. Unframed patterns are used on T1 repeatered lines, but framed patterns are necessary if the facility traverses other types of transmission equipment. Timing recovery difficulties may result if a DS1 pulse stream contains more than 15 consecutive zeros or if the ones density is less than 12.5%. Equipment interfaces within a DS1 facility may be wrongly conditioned for AMI or B8ZS operation, resulting in excessive bit errors. These problems may be identified by using the appropriate stress test patterns.

The QRSS pattern was used originally for testing facilities to be used for speech traffic, but it is still employed to check the basic performance of a path. It checks the timing recovery because it contains strings of up to 14 zeros and includes low ones-density sequences.

The 3-in-24 pattern contains 15 consecutive zeros and has the minimum allowable 12.5% ones density, which stresses AMI-conditioned facilities to the normal limits. If a piece of equipment incorrectly conditioned for B8ZS is present in an AMI facility, the unwanted B8ZS substitution will cause errors in the signal received at the distant end. If the 2-in-8 test pattern is then substituted for the 3-in-24 pattern, the error performance will be restored to normal because the 2-in-8 signal contains no more than four consecutive zeros and cannot cause B8ZS substitutions.

Table 11.1
Stress-testing Pattern

Pattern	Description
QRSS	Quasirandom stream consisting of combinations of 20 bit words with no more than 14 consecutive zeros.
3 in 24	010001000000000000000100... May be framed or unframed. Contains 15 consecutive zeros, with 12.5% ones density.
1 in 8	010000000100000001000000... May be framed or unframed. Only one bit position in each 8 bit sequence is occupied by a one.
All ones	May be framed or unframed.
All zeros	Pattern has to be coded with B8ZS or ZBTSI before being sent into line.
2 in 8	0100001001000010010000010... May be framed or unframed. Contains 4 consecutive zeros.

For testing B8ZS-conditioned facilities, the 1-in-8 pattern provides the minimum ones density, but without causing B8ZS substitutions along the path because the sequences of zeros are limited to seven. If an equipment-conditioning error has placed an AMI element in a B8ZS facility, the facility will show errors at the receiving end. This condition can be verified by sending an all-zeros pattern coded for B8ZS into the line, which will cause a complete loss of data or an incorrect pattern to appear at the receiver.

The all-ones pattern is used to stress the regenerators in a T1 line. This pattern causes the power consumption of the regenerators to increase to its maximum value, and any limitations due to power feeding over the cable system become apparent.

11.7 SONET SYSTEMS

11.7.1 General

The operations aspect of SONET systems, which is much more sophisticated than the corresponding functions in other digital transmission systems, makes use of the large overhead capacity of SONET. The details of the SONET operations, administration, maintenance and provisioning (OAM&P) are still evolving, and the standards are in preparation [1].

11.7.2 Operations System Support

The operations systems support (OSS) for SONET uses data links from each network element (NE) to a control center, either directly or via intermediate collection points. All of the data available at any NE is available to the system operations personnel, and by this means a global view of a SONET network can be compiled. Commands can be sent directly to an NE, and the resulting action confirmed from the messages sent back from the affected NE sites. Spontaneous reports of status and alarm conditions are sent from the NE locations to the control center, and it can be arranged that such reports are transmitted at fixed intervals. Security of access to the systems is provided by means of passwords and individual identification codes. Various levels of access security and classes of commands are defined, so an individual has access to only those commands required to perform the assigned tasks.

One of the major standardization efforts concerns the format and terminology of the messages exchanged between the NE sites, and between the NE sites and the control center. SONET systems are being designed and manufactured by a number of vendors in North America and overseas, and initially each system is using proprietary formats for the messages—even if they use a common syntax for message construction.

11.7.3 Local OS Access

All NE equipment has a provision for local access and may use a *dumb terminal*, which uses software in the NE, or a personal computer (PC). The availability of portable notebook computers now makes it feasible to provide these to personnel visiting the sites. From any site on a network at least some segment of a system may be addressed, assessments of the system conditions obtained, and necessary corrective measures implemented. The same security considerations that apply to control-center access apply to local access.

11.7.4 Alarm Surveillance

Alarms due to equipment failure and problems with the SONET line signals are indicated at the sites by lamps on the bays and visual and audible office alarms. Most sites are not attended on a full-time basis, so the alarm conditions are also sent to the appropriate major site for attention by the operating and maintenance staff. Alarms may be service-affecting or nonservice-affecting and are divided into critical, major, and minor alarm categories. The analysis of the indications from a number of sites enables the location and nature of the problem to be determined.

11.7.5 Performance Monitoring

The SONET systems provide in-service performance monitoring, which allows any degradation in performance to be detected before it deteriorates to unacceptable levels. The parameters are monitored at optical, section, line, and path levels, which facilitates tracing the source of the problem to a particular NE.

At the optical level, the laser bias current and the transmitted and received optical power levels are monitored, and any deterioration of laser or detector performance can be observed. The parameters monitored separately at the section, line, and path levels typically include coding violations, errored seconds, severely errored seconds, unavailable seconds, and degraded minutes. Each of these parameters is stored in registers, and the accumulated values for 15-minute and, in some cases, one-hour intervals and the totals for 24-hour intervals may be sent as reports to the operations center. A threshold value may be established for any parameter for the 15-minute or 24-hour interval, and a threshold crossing alert signaled if this value is exceeded. Performance-monitoring data may be requested from any NE, and the threshold values may be changed locally or remotely.

11.7.6 Provisioning

The SONET system has a variety of selectable parameters, which are provisioned by software commands. Once the equipment is installed and properly fitted with plug-in modules, there is very little, if any, manual site provisioning required to place it in service. The low-speed DS1 and DS3 interfaces, the high-speed optical line and tributary interfaces, the protection switching parameters, the desired performance-monitoring surveillance, and other features can be provisioned and checked remotely. Remote provisioning may be carried out from another equipment site or via the OSS from the control center. A complete end-to-end system may have all of its parameters at every site provisioned, the system tested, its performance monitored, and the system placed in service without any personnel being needed at the equipment sites.

11.8 OPERATIONS SUPPORT SYSTEMS

11.8.1 General

As a network grows in size, manual control becomes increasingly impractical, and a centralized operations support system (OSS) approach becomes necessary. Each type of DCS and SONET system incorporates its own operations support (OS) functions, and the details of these vary from vendor to vendor. In addition, the management of the testing capabilities that are integral to the network requires its own OS. The control center for a network consisting of a number of DCS 1/0, DCS 3/1, and DCS 3/3 systems located at many sites and probably also including SONET transmission systems requires a large number of control terminals to handle the operation of all of these systems. The operational complexity is increased when systems from more than one vendor are deployed in the network. This problem may be overcome by using only one vendor for a particular type of system, but this single-source procurement leaves the network owner completely dependent on that vendor for continued equipment supply and technical support of the product.

11.8.2 OSS Integration

It is a desirable objective to reduce the number of different OS systems at the control center. One way to do this is by making one OS system control more than one type of DCS or SONET system. A major difficulty encountered in attempting OSS integration is incompatibility of message formats, due to the variety of language, syntax, and vocabulary used in the systems, as discussed in Section 11.4. Another factor to be considered is that some operational functions are performed only by one type of system. For example, private-line circuit provisioning is a major application of the DCS 1/0 systems; the DCS 3/3 is involved in network survivability at the DS3 level; and provisioning of DS1 facilities is performed using the DCS 3/1. There are, therefore, advantages to having separate OS terminals and operators dedicated to performing particular tasks such as private-line provisioning.

It will be a difficult and time-consuming task to reach an agreement on standards for OS functions for all DCS and SONET systems. In the meantime, proprietary OS functions will be provided by the system vendors. OS integration will eventually allow a reduction in the number of control terminals needed for those functions that are common to a number of systems, such as performance monitoring and alarm surveillance.

11.8.3 OSS Functional Requirements

Some basic OSS functions are common to all of the network elements. These include the real-time surveillance of alarms and changes of status, the updating of the network databases, the notification of network problems requiring corrective action, the ability to send commands to the network elements and receive messages from them, and the ability to gain authorized access to the network elements using a flexible security scheme.

Enhanced functions, some of which apply only to particular network elements, include the setting of optional parameters; the collection of performance-monitoring data; the setting of performance thresholds, configuration, and management of circuits and facilities; the setting of conference and broadcast circuits; maintenance and diagnostic activities; inventory management; and testing management. The management of circuits and facilities in a large network is a critical function that requires the assignment databases to be kept completely up-to-date and accurate, so end-to-end circuit routing can be tracked.

11.8.4 Partitioning

Partitioning capability is important in the creation, access, and control of virtual private networks. By this means a portion of the capacity of a network element is assigned to a particular customer, who then controls the access to it and is able to configure and administer it. For example, a number of the transmission ports on a DCS system may be allocated to a customer, who may connect, disconnect, and rearrange DS1 or DS3 facilities; send and receive messages; and receive notification of alarms without any other user of the DCS having access to any of this information.

In addition to the capability of providing virtual private networks, partitioning allows the division of the network into separate pools of circuits, which may be assigned as the responsibility of different operational groups. An example of this is the allocation of private-line DS3 facilities to a private-line group, while the management of the remaining DS3 facilities is allocated to another operations group.

11.8.5 Graphical User Interface

A graphical user interface (GUI) allows the creation of real-time graphic representations of the network and its network elements. The GUI capabilities should include the ability to build graphic representations of networks, event notification via graphic display, and screen zoom for a detailed view of a portion of the network.

The GUI has to allow for hierarchical progression in the presentation of information. From a view of the entire network, the operator must be able to select a view of a site, an equipment bay, an equipment shelf, an individual plug-in module, and the designation of a lighted alarm LED on the module. The operator must be able to change from any level of detail to another as required and select for retrieval and display performance-monitoring data and other critical operational information. The procedure for using the GUI has to allow the operator to change from one display to another without continual reference to the instruction manual.

11.9 OSS ARCHITECTURE

The provision of an OSS that incorporates all of the features described in the previous sections is a complicated task. Each network element vendor would like to supply its own proprietary OSS but is aware that it is a system with a very limited market size. Each network operator would like to procure an OSS tailored to meet the needs of that particular network, with as many custom built features as possible. For these reasons the OSS architecture is discussed here in only general terms.

A possible architecture is shown in Figure 11.1. The top of the figure indicates a software module for management of the network facilities database. Different software application modules are used to manage the different classes of network element, such as the DCS 1/0, the DCS 3/1, the DCS 3/3, and the SONET equipment. Other management functions include controlling and performing tests through the test management module. The DS1, DS3, and other performance-monitoring functions are placed under the performance-monitoring module. Additional management functions can be loaded into the computer platform as needed in the future.

Communication connection to the network is managed by the network communications interface. Increasingly, subnetwork operational controllers are being deployed. These controllers, more commonly known as mediation devices, serve the function of a translator between the network elements and the management system, thus allowing more flexibility in the development of the management application modules. In addition, the controllers may also serve as concentrators for multiple network elements of the same class.

The DCS and SONET systems are connected to the appropriate segment of one of the subnetwork operational controllers, which are located at major sites on the network. These controllers are, in turn, connected to a packet data network, to which the network management system is also connected.

Redundancy of the OSS is of critical importance because a failure could cause the loss of all of the network-management features. Hardware duplication is one method of providing redundancy, but this is being superseded by synchronous distributed computing. Flexibility is enhanced by being able to load several management modules into one computer or to load one module into each of several computers. Reliability is achieved by using multiple computers connected to a local area network (LAN) or to a wide area network (WAN). The database of multiple central processing units (CPU) can be synchronized to

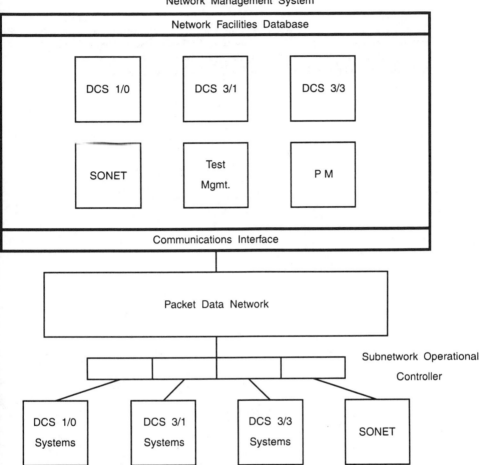

Figure 11.1 OSS architecture.

within 10 ms of the master processor. This distributed computing allows the platforms to be physically dispersed for safety and the LANs can be interconnected through high-speed data circuits.

Network management systems based on distributed computing provide flexibility for the inevitable changes that will result from future growth, and they provide the essential redundancy. Future needs for increased capacity or capability can be met by upgrading to bigger computers or adding more machines to the distributed computing system.

References

[1]   ANSI T1.214-1990, American National Standard for Telecommunications, Operations, Administration, Maintenance, and Provisioning (OAM&P)—Generic Network Model for Interfaces between Operations Systems and Network Elements.

[2]   Bellcore, Technical Reference TA-TSY-000396, Transaction Language 1.

Survivability with DCS 3/3 **12**

12.1 INTRODUCTION

Major failures in public communications networks cause havoc and receive widespread media attention. Not only are people unable to make or receive telephone, data, and facsimile calls, but vital services such as air-traffic control links may be put out of action. Similar failures in private networks may be disastrous for the company involved and will receive considerable adverse attention from corporate management. Some catastrophic failures in the public network sector have caused government agencies to initiate studies of survivability techniques to prevent such breakdowns from occurring.

The basis of survivability is to ensure that when a failure occurs in a network, the interrupted traffic is rerouted to its destination over an alternate route unaffected by the failure. It may not be necessary to reroute all of the traffic, depending on the priority assigned to various classes of service. For example, in switched telephone service sufficient intermachine trunks (IMTs) are provided between the switches to carry the traffic at the busiest times of day. If some of these IMTs are lost the effect will be minimal for much of the time, and the deterioration of service in the busy hours can be tolerated until the network is restored.

12.2 ROUTE DIVERSITY

One method of reducing the risk of losing all service between two points on a network is to provide route diversity. A commonly used planning rule is that no more than 50% of the traffic between any two points on a network is to be carried over the same physical route. Figure 12.1. shows a network segment with three sites: A, B, and C. In this case, the traffic between A and B is divided so that half is routed directly from A to B and the other half is routed

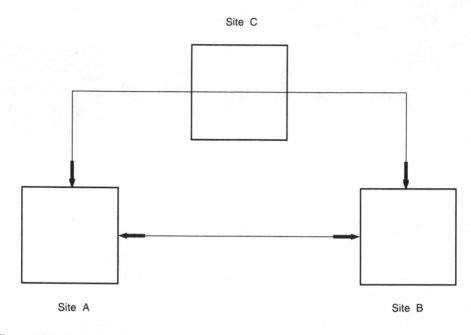

Figure 12.1 Route diversity.

from A to C to B. Similarly, the traffic between A and C is divided between the direct A-to-C link and the routing from A to B to C. One of the disadvantages incurred in providing survivability by route diversity is that in a very large network, this may result in a direct route between two points of only 100 miles and a diversity route of 3,000 miles. This technique can be used for traffic similar to the switch IMT circuits, where up to half of the circuits can be allowed to be interrupted by a failure. High-priority circuits, such as air-traffic control links, may have to be duplicated over two diverse routes so service is maintained if one route is lost.

12.3 ACCESS SURVIVABILITY

Route diversity is suitable for traffic that is carried between two points connected by two or more routes. It cannot be used where one of the sites is located on the link between major points on the network. Traffic that originates and terminates at a local telephone company office gains access to the network at these intermediate sites, so in this case there are no alternative transmission paths. Therefore, no form of route diversity can be used for the survivability of this access traffic. In Figure 12.2, three access sites, X, Y, and Z, are in the link

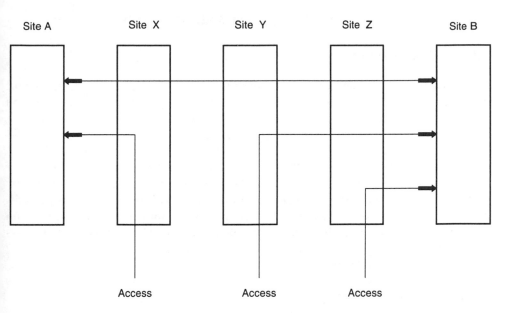

Figure 12.2 Access traffic homing.

between the major sites A and B. For descriptive purposes, assume that all of the traffic from X homes on site A, and traffic from both Y and Z homes on site B. Homing means that all of the traffic from an access site is taken to one major site, where it either terminates or is sorted and then passed on to the assigned route to its final destination. In the event of a break in the path between X and A, such as a cut in a fiber-optic cable, all of the traffic to and from site X is lost. The immediate effect is loss of revenue to the network operator, but the longer term result may be loss of customers.

The objective of a survivability plan for access traffic is to restore traffic as rapidly as possible after a path failure. All of the switched traffic circuits, and as many of the other service circuits as possible depending on their priorities, are rerouted. Until the advent of the DCS 3/3, the rerouting to alternate paths was carried out by manual patching on DSX-3 frames. This is a process that may take hours and that has to be reversed when the break is repaired. Where large numbers of DS3 facilities are involved, errors in the patching are almost inevitable. A more rapid and controllable approach is to provide a number of loops on the transmission network and to reverse the direction of flow of the broken facilities around the loop when a path cut occurs. The normal facility routing is restored after the system break is repaired. The next section describes the restoration process when the DCS 3/3 switch is used in conjunction with the reverse direction protection switch (RDPS).

12.4   RESTORATION USING DCS 3/3 AND THE RDPS

12.4.1   Provision of Loops

Rerouting of DS3 access traffic using remotely controlled DCS 3/3 systems and network loops is much more efficient than manual patching and reduces the time required to restore service from hours to minutes. A typical network with multiple loops is illustrated in Figure 12.3. In most cases, more than one DS3 needs to be restored at each access site, and the available DS3 restoration capacity on each segment of the loop has to equal or exceed the total number of DS3 facilities to be rerouted due to the failure. In addition, many of the transmission links are common to more than one loop, and if there are multiple failures requiring the use of the same link, DS3 capacity may allow only one of the restoration loops to be activated. Access at the DS3 level to all of the transmission system capacity, including working and protection channels, is necessary at all the sites on the loops.

12.4.2   Normal Conditions

The configuration of a loop with sites A, B, C, and X under normal conditions is shown in Figure 12.4. The arrangement of relays and other equipment at site X comprises the RDPS, and this is used in conjunction with DCS 3/3

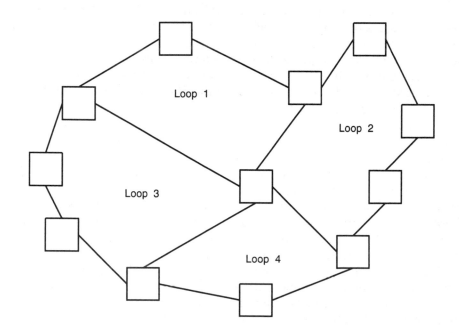

Figure 12.3 Network loops.

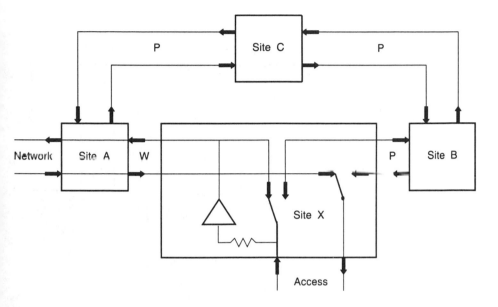

Figure 12.4 Normal conditions.

systems at the node sites A, B, and C to affect the traffic reversal around the loop. This is a simplified portrayal of the arrangements and is intended only to illustrate the principles employed. The normal access traffic path is through the closed contacts of the RDPS, over the working transmission system path W to site A and via the DCS 3/3 to its destination. The bridging repeater at site X is bypassed in this situation. The protection paths P, which will be used to complete the loop reversal, are shown but are not cross-connected at sites A, B, and C.

12.4.3 Rerouting in Failure Mode

When a cable break occurs between site A and site X, the access traffic from X is interrupted, as shown in Figure 12.5. The RDPS at X is then operated, connecting the access traffic over the protection path P to site B. Cross connections are made on the DCS 3/3 switches at A, B, and C, so that the traffic is once more connected to its destination through site A. Service will remain on the reversed loop path until the break is repaired, and the path between A and X is once again available for service. The bridging amplifier at site X now applies the transmit access signals towards site A, on the broken *working* path W. The DCS 3/3 at site A also has a bridging connection onto the broken path, applying the transmit signals towards site X. These bridging signals will be used later during the restoration of the system to normal conditions.

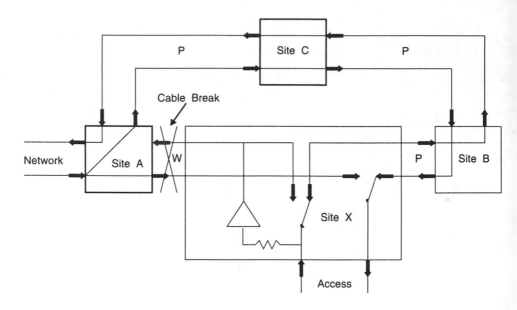

Figure 12.5 Cable break.

12.4.4 Restoration to Normal

The restoration of the traffic to its normal flow, after the broken links are re-paired, has to be carried out with the minimum of interruption of service. The original disruption following the break is unavoidable, but to lose service again just to restore the original network conditions for the service provider's benefit leads to justified customer complaints. A break of less than 50 ms does not cause problems in telephone service, although data service loses some bytes that must be retransmitted. Various techniques have been considered to syn-chronize the switching operation of the DCS 3/3 switches and the RDPS to re-store both directions of transmission simultaneously within the 50 ms limit. No satisfactory method has been evolved, even with the use of precision clocks at each site. However it has been found unnecessary for both directions of transmission to be moved back to the normal routing simultaneously, as long as there is not a long interval between the restoral of the two directions. The important factor is that neither direction of transmission should be interrupted for longer than 50 ms during the restoral cycle.

After the path break has been repaired and continuity between sites A and X restored, the return to normal conditions is initiated. Figure 12.6 illustrates the restoral of the receive direction at site A. The RDPS remains in the reverse-loop setting, and the traffic transmitted from A to X is still carried on the protection path. However, the access traffic now flows via the bridging amplifier over the working system and through the DCS 3/3 to its original destination. This traffic

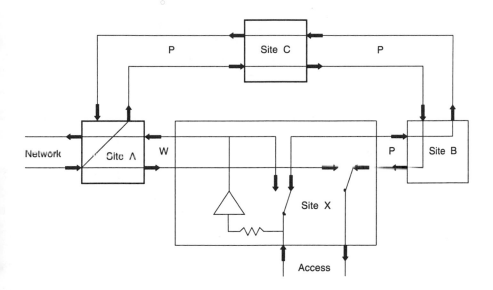

Figure 12.6 X–A traffic restored.

is also still being carried over the protection system back to A, but it is no longer connected through the DCS 3/3. The operation of the RDPS back to its original setting restores the site X receive traffic to its normal path, as illustrated in Figure 12.7. The final step is to remove the surplus DCS 3/3 connections to restore the conditions of Figure 12.4.

12.5 MAPS AND PLANS

The restoration process described in the previous section uses various cross connections at the DCS 3/3 switches, as illustrated in Figures 12.4 through 12.7. As already mentioned, these diagrams show only one DS3 facility involved in the restoration, but in actual networks it is usual for several of these facilities to be involved. It is possible, when a link failure occurs, to determine from the network database which cross connections have to be added, deleted, or changed for each DS3 facility, at each of the DCS 3/3 systems, at each stage of the restoration process. Each of these changes could be made by individual commands from the central control site, but this would be a lengthy task and might result in some incorrect connections due to operator error. Because it is important to restore service as rapidly as possible, preplanning of the cross connections required at the DCS 3/3 systems for each possible failure scenario is necessary.

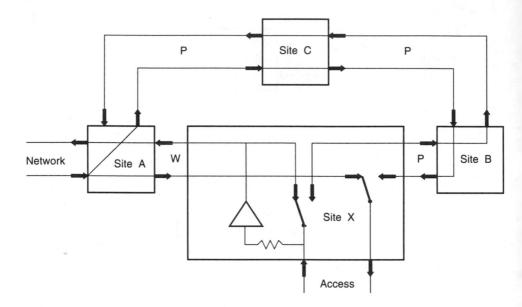

Figure 12.7 X–A and A–X traffic restored.

The concept of the cross-connection map was introduced at the end of Chapter 10. One description of a map is that it comprises the list of all of the DS3 cross connections that are required to be in place in the DCS 3/3 switch when that map is invoked. In this definition no one map has any knowledge of the existence of any other map. When the change is made from one map to another, any cross connection that already exists and is needed again in the new map is left intact, rather than being disconnected and reconnected. For any particular DCS 3/3, a single command is sent to change to a new map, e.g. "swap to map 79." Each DCS 3/3 is capable of storing at least 100 maps, so all restoration situations are likely to be covered. An alternative approach defines a map as the list of cross connections to be added, deleted, or changed to go from one DCS 3/3 configuration to another, e.g. from "normal" to "break between sites A and X." Both of these methods are available in DCS 3/3 systems.

Because multiple DCS 3/3 systems are involved in every restoration scenario, a set of plans is evolved showing the maps to be invoked at each DCS 3/3 for each possible network link break that can occur. Generally, only a few of the DCS 3/3 systems are involved in the restoration due to any one path break, so the plans are not unduly complicated. The normal condition of the network when there are no breaks is reflected in a *baseline plan*, which lists the *baseline map* for each DCS 3/3. When a break occurs, the operations-center personnel determine the location of the break and invoke the appropriate restoration plan by sending the map change commands to the affected DCS 3/3 switches and the changeover commands to the RDPSs involved. After the system is repaired, the final step is to return the network configuration to the baseline plan, in preparation for the next break.

Networks evolve rapidly, and DS3 facilities are added, deleted, and changed as service requirements dictate. It is essential that the maps and plans are kept up to date to reflect these events, or the invocation of restoration plans will not restore the facilities correctly, which results in confusion and lost revenues.

12.6 SURVIVABILITY WITH SONET RINGS

Restoration of traffic following a cable break, using the DCS 3/3 and RDPS technique, reduces the out-of-service period from hours to minutes. The advent of SONET systems makes possible service restoration in 50 to 100 ms, using a self-healing ring method, a technique described in Chapter 14.

SONET 13

13.1 THE SYNCHRONOUS OPTICAL NETWORK

The synchronous optical network, abbreviated to SONET, is a transmission system technology for use in fiber-optic cable networks [1]. It is *not* a telecommunications service. It is anticipated that SONET systems will progressively replace the existing point-to-point asynchronous systems, resulting eventually in completely synchronous networks. Networks using asynchronous systems have numerous disadvantages that are overcome in synchronous networks, and these are discussed in the following sections.

13.2 DISADVANTAGES OF ASYNCHRONOUS SYSTEMS

A number of vendors supply equipment for fiber-optic cable systems, but there is no standardization of the system parameters. Each of these asynchronous systems uses a different format for the optical signals on the fiber cable, and each format is proprietary to the manufacturer of the system. In addition, there is no standardization of the line bit rates or of the operation of the line protection switching, so any one end-to-end system must be implemented with all of its network elements supplied by the same vendor. In addition to these problems of incompatibility, the asynchronous nature of these systems makes it impossible to add and drop traffic at an intermediate point on the system without demodulating the entire bit stream to the DS3 level and then remodulating it.

13.3 ADVANTAGES OF SYNCHRONOUS SYSTEMS

The SONET system has been designed to meet a set of standards, which include the format of the line signals, a hierarchy of synchronous bit rates, protection switching operation, and substantial overhead capacity.

These features allow equipment from multiple vendors to be operated on a point-to-point system and the interconnection of systems on an optical basis without demodulation to DS3 level. The synchronous nature of the line signals allows the dropping and inserting of traffic at intermediate points on a system using an add/drop multiplexer (ADM), instead of the back-to-back terminals used in asynchronous systems [2]. The overhead capacity is available for use with a comprehensive operations support system (OSS) and allows access at local and remote sites to alarm and status information and performance-monitoring data. Provisioning of any of the network elements may be performed remotely using software control from network sites or from a central control site.

The SONET systems accept the asynchronous DS1 and DS3 traffic that will continue to exist in networks due to the vast amount of asynchronous terminating equipment, such as the M1-3 multiplexer, already deployed. In addition to DS1 and DS3, SONET systems are capable of handling broadband traffic such as high-speed data and the future high-definition TV signals.

The SONET technique makes possible completely synchronous networks, using the interconnections required by the network traffic, instead of a collection of point-to-point systems interconnected at the DS3 level at terminal sites. The SONET network will be remotely software controlled using a sophisticated OSS. The result will be networks using reduced amounts of equipment, providing better performance and lower maintenance labor costs. The remote performance monitoring and alarm/status information features will reduce the duration and cost of repairing system interruptions.

In Europe, a similar but different standard for synchronous systems has evolved, called the synchronous digital hierarchy (SDH). The SONET and SDH standards allow international traffic to be interconnected between these two types of synchronous networks.

13.4 BASIC SONET TECHNOLOGY

This section covers the SONET technology in sufficient depth to explain the operation and features of the systems. The details of the various frame structures, the functions of individual overhead bytes, and the ways in which various payloads are handled in the SONET multiplexing process are described in the ANSI and Bellcore publications [1,3]. The SONET standards are still evolving, particularly in the OSS area and in the area of the functions that are used in the SONET rings described in the next chapter.

Any attempt to include these topics in great detail here would not be productive, as much of the information would become obsolete very quickly. However, the basic information covered here is well-established and unlikely to be changed to any significant extent.

13.4.1 Section, Line, and Path Overheads

In addition to the digital traffic carried on a SONET system, which is called the payload, the line bit stream includes blocks of overhead bytes defined in relation to the system network elements. Three types of overheads are used called the section, line, and path overheads. The section and line overheads combined are called the transport overhead. Figure 13.1 shows the elements of a SONET system and the originating and terminating points of each type of overhead. A regenerator is a section terminating equipment (STE) only, while the terminal and the ADM function as both a section and a line terminating equipment (LTE). At the end of the system, a network element that converts the asynchronous DS1 or DS3 traffic into SONET formats and back again functions as a section, line, and path terminating equipment (PTE). The section overhead originates at the output of an STE and terminates at the input of the next STE on the cable. The line overhead is generated at the output of an LTE, traverses any intermediate STEs unchanged, and terminates at the next LTE. The path overhead originates in the PTE, where it combines with the asynchronous traffic and remains unmodified until it reaches the distant PTE, where the asynchronous signals are removed from the STS-1 structure.

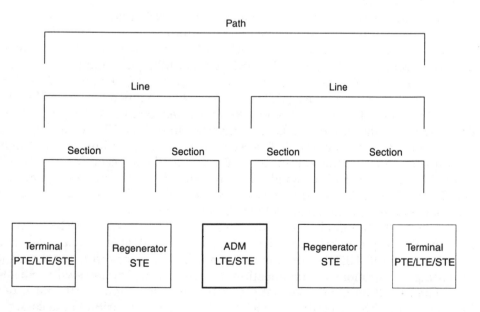

Figure 13.1 SONET overheads.

13.4.2 Multiplexing Techniques

Asynchronous multiplexing, described in Chapter 3, is based on the 1.544 Mbps DS1 bit stream. Four DS1 signals are combined to form one DS2 at a speed of 6.312 Mbps. The DS2 rate is higher than four times the DS1 rate to allow additional bits for bit stuffing purposes. This is necessary because none of these signals is synchronized, and variations in the speeds of the DS1 bit stream have to be accommodated. The same type of multiplexing takes place in the DS2-to-DS3 step, with the result that no signal in the hierarchy has a speed that is an exact multiple of a lower bit rate.

In synchronous multiplexing, all of the signals are derived from precision sources, and the bit rates at each level of the hierarchy are exact multiples of the bit rate of the basic SONET signal. This is called the synchronous transport signal level 1 (STS-1) and has a bit rate of 51.84 Mbps. The STS-1 is the building block of the SONET hierarchy, and the higher speed signals are formed by multiplexing a number of STS-1 signals to form an STS-N signal, where N is an integer, so that any higher speed signal is an exact multiple of 51.84 Mbps. The demultiplexing process separates the STS-N stream into N STS-1 signals, with no bits left over, since no stuffing is employed in the multiplexing process. This technique, where each bit rate in the hierarchy is an exact multiple of the basic rate, is one of the major features that distinguish SONET from asynchronous fiber-optic systems.

The multiplexing of the input signals into SONET optical line blocks is illustrated in Figure 13.2. Asynchronous DS3 traffic at the 44.736 Mbps rate is combined with stuffing and other overhead bits and mapped into the payload of a 51.84 Mbps STS-1 signal. The DS1 at 1.544 Mbps is combined with stuffing and overhead bits into a virtual tributary 1.5 (VT1.5) signal at 1.728 Mbps. A number of virtual tributary (VT) signals are then mapped into an STS-1 payload. The terminals can also accept externally generated STS-1 signals. In the OC-3 terminal, the input signals are converted into three STS-1 blocks, which are multiplexed into one STS-3. The STS-3 is then changed into an OC-3 signal for application to the fiber-optic cable. The terminal can accept 84 DS1 blocks, three DS3 or three STS-1 signals, or any mix of these types of signals that equals one STS-3. A metallic interface has been defined for the STS-1 but has not been widely implemented on SONET equipment.

The OC-12 terminal accepts the same input signals as the OC-3 terminal, and an OC-3 input interface is also available. An optical interface at the low-speed side of a SONET terminal is called a tributary port, and is the means for providing synchronous interconnection between SONET systems within a site.

The OC-48 terminal does not generally offer DS1 input ports, but it does have DS3 and STS-1 metallic and OC-3 and OC-12 optical tributary access.

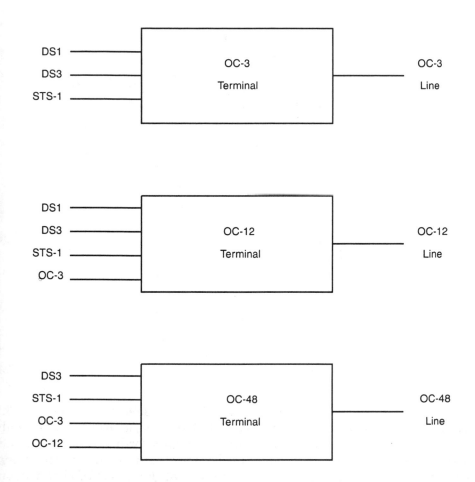

Figure 13.2 SONET multiplexing.

13.4.3 Optical Line Rates

The STS-N streams are the electrical signals generated by the multiplexing process. These are converted to optical signals for application to the fiber-optic cable and changed back to electrical signals before the demultiplexing process. The optical signals have exactly the same rates as their electrical counterparts, and are designated optical carrier level N (OC-N). For both the electrical and optical signals, the hierarchical bit speeds are all multiples of the basic 51.84-Mbps rate. While the OC-N line signals could use any integer value for N, Table 13.1 shows the values that have been standardized, the corresponding line rates, and the voice channel capacity [3].

Table 13.1
Optical Line Rates

OC Level	Line Rate (Mbps)	Voice Channel Capacity
1	51.84	672
3	155.52	2,016
9	466.56	6,048
12	622.08	8,064
18	933.12	12,096
24	1,244.16	16,128
36	1,866.24	24,192
48	2,488.32	32,256
96	4,976.64	64,512
192	9,953.28	129,024

The line rates commonly used in the SONET systems being deployed in networks are OC-12 and OC-48, but some OC-3 systems are used on links with low traffic loading. Techniques are being developed for practical OC-96 and OC-192 systems, and these will doubtless be introduced in the next few years. There is obviously concern about carrying tens of thousands of channels on one pair of fibers, due to the catastrophic effects on the network of either fiber or equipment failures. This concern may limit the use of systems having the highest attainable SONET rates and will certainly make survivability and alternate routing techniques very important network considerations.

13.5 STS STRUCTURE

13.5.1 Composition of the STS-1 Frame

The STS-1 bit stream is divided into frames in the same way as the DS1 and DS3 signals previously described. The STS-1 frame is timed from a precision clock source, and each frame is exactly 125 microseconds long, so there are 8,000 frames in each second. The frames consists of 810 bytes of eight bits each, making the bit rate (8,000 frames/sec) × (810 bytes/frame) × (8 bits/byte) = 51.84 Mbps.

The STS-1 frame contains the section overhead (SOH), the line overhead (LOH), the path overhead (POH), and the payload bytes, as shown in Figure 13.3. The frame is shown as a block with 90 columns nine rows deep with 1-byte cells. The STS-1 frame is transmitted starting with all the bytes in row 1, counting from the left, followed by the bytes in row 2, and then the bytes in each successive row until all 810 have been transmitted.

The SOH occupies the first three bytes of the first three columns, and the LOH occupies the remaining six bytes of these columns. Since the POH and LOH together comprise the transport overhead, it can be seen that the first three columns of the frame, consisting of 27 bytes, are allocated to the transport overhead function. The fourth column of nine bytes is assigned to the POH function, and the remaining 86 columns provide the payload capacity of the frame. The POH and the payload together are known as the synchronous payload envelope (SPE). The SPE capacity occupies 87 columns and accommodates 783 bytes.

The allocation of functions to the frame byte positions is the same for every STS-1 frame. The information in the payload bytes and the bit values in the overhead bytes are the items that vary from frame to frame.

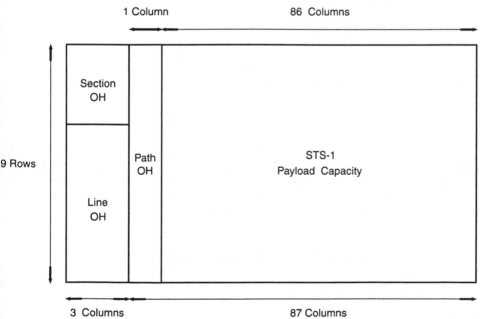

Figure 13.3 STS-1 frame.

13.5.2   STS-1 Overheads

The principal features provided by the bytes contained in the SOH, LOH, and POH are listed below.

13.5.2.1   Section Overhead (SOH)

The A1 and A2 bytes carry the framing information for the STS-1 frame, and B1 carries performance data at the section level. The section orderwire, accessible at all sites including regenerators, is carried in byte E1, and a similar channel assigned to the system user uses byte F1. Bytes D1 through D3 form a 192-Kbps section data communications channel (DCC).

13.5.2.2   Line Overhead (LOH)

The K1 and K2 bytes are used for the automatic protection switching (APS) signaling channel, and bytes H1 through H3 contain the STS pointer information, which is explained later. Byte B2 carries performance data for the line level. The express orderwire channel occupies the E2 byte, and the 576-Kbps express DCC uses bytes D4 through D12.

13.5.2.3   Path Overhead (POH)

A system user channel is carried in byte F2; byte B3 carries performance data; and path status information occupies byte G1.

13.5.3   STS-N Frame Format

The format of an STS-N signal is shown in Figure 13.4. The transport overhead, consisting of the section and line overhead, now occupies $3 \times N$ bytes in each of the nine rows for a total of $27 \times N$ bytes. The STS-N SPE capacity is the remaining $87 \times N$ bytes in each row for a total of $783 \times N$ bytes. The frame length is 125 microseconds for every value of N.

13.6   SYNCHRONOUS PAYLOAD ENVELOPE

13.6.1   General

The STS-1 SPE consists of the payload and the POH and occupies 783 bytes. The types of traffic to be carried over SONET systems include DS3 at 44.736 Mbps, DS1 at 1.544 Mbps, DS1C at 3.152 Mbps, DS2 at 6.312 Mbps, and the CEPT E1 basic signal at 2.048 Mbps where international traffic is involved.

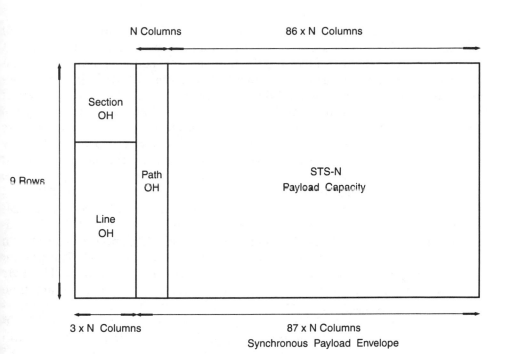

Figure 13.4 STS-N frame.

One DS3 is accommodated in an SPE, but for the lower rate traffic, multiple signals can be fitted into the SPE by using an intermediate modulation stage called a virtual tributary (VT).

Whichever type of traffic is being carried, the STS-1 SPE always contains 783 bytes and is timed from the same precision clock as the STS-1 frame. As far as the SONET line transmission system is concerned it sees only the STS-1, and the type of traffic being carried is invisible.

13.6.2 Floating SPE

The SPE is carried intact through a SONET network, from the point of origination of the traffic to the point where the SPE is broken down, and the traffic is delivered to its destination. Along its path through the network, the SPE may be carried in STS-1 frames on a succession of transport links. The transport overhead terminates at the input side of network elements, such as an ADM, and is created again at the output side. For this reason, and to allow for small timing differences in a synchronous network, the SPE is allowed to float anywhere within the SPE capacity of the STS-1 frame.

Figure 13.5. shows a succession of three STS-1 frames and the positions of their associated SPE blocks. Of the 90 columns in each frame, the transport overhead occupies the first three, and the remaining 87 columns form the SPE capacity. Each SPE travels partly in its own STS-1 frame and partly in the following frame. This is the normal condition because very rarely does the first byte of the SPE occupy the first position in the STS-1 frame SPE capacity. During the evolution of the SONET standard there was a proposal to use storage registers to enable the SPE to be aligned exactly with the frame SPE capacity. The very large amount of storage that would have to be provided and the delay that would be introduced into the transmission of the STS-1 frame caused this approach to be abandoned, and a technique using STS pointers was adopted instead.

The pointer is used to indicate which byte position in the STS-1 SPE capacity is occupied by the first byte of the SPE. The numbering of the SPE capacity byte positions is shown in Figure 13.6. The H1, H2, and H3 bytes shown in the transport overhead area are the pointer bytes of the STS-1. The pointer bytes always occupy the same positions in the transport overhead. The byte position directly following H3 is designated byte 0. The positions are then numbered successively across each row, until byte position 782, which immediately precedes byte H1 of the next STS-1 frame. The pointer value is the offset between byte position 0 and the actual position of the first SPE byte, and it is carried as a binary number using four bits of byte H1 and six bits of byte H2.

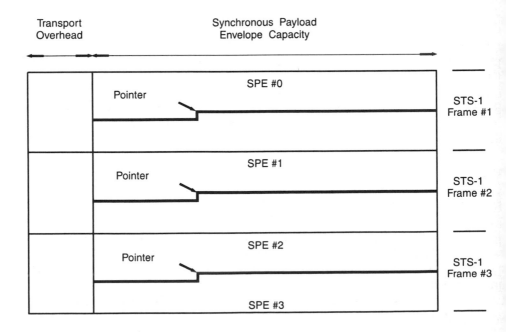

Figure 13.5 Floating SPE.

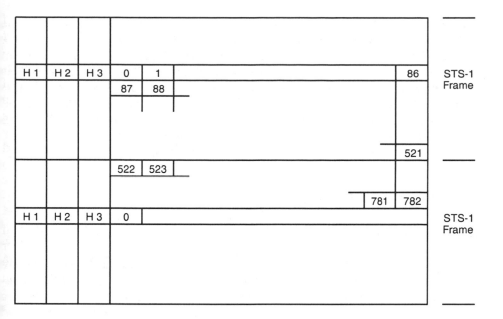

Figure 13.6 Pointer numbering.

Because the bit rates of the STS-1 frame and the SPE can have very small differences, the SPE will occasionally have one byte too many, or one too few, to fit exactly into the frame. If the SPE has one extra byte, this byte is placed in the H3 location in the transport overhead. When the SPE has one too few bytes, dummy information is placed in the first byte position of the SPE capacity. These conditions are noted by changing the values of the first four bits of byte H1.

13.6.3 DS3 Mapping Into SPE

The asynchronous DS3 signal is the only traffic block mapped into the STS-1 without the use of a VT stage of multiplexing. The payload byte capacity of the STS-1 frame consists of 86 columns of nine rows each, each byte having eight bits. The frame duration is 125 microseconds, corresponding to 8,000 frames per second. The payload capacity bit rate is $86 \times 9 \times 8 \times 8,000 = 49.936$ Mbps. The mapping technique uses a combination of fixed stuffing bits, stuffing opportunity bits, and other overhead bits to fit the 44.736 Mbps DS3 into the payload capacity.

In Section 13.3, the ability to drop and insert individual DS3 blocks from a line, without demultiplexing the complete line signal, is listed as a major attribute of SONET systems. A DS3 mapped into an SPE meets this requirement because any STS-1 frame and its associated SPE can always be located in the OC-N line signal. For those DS3 streams that contain 28 DS1 channels as the payload, the DS1 signals are also mapped into the SPE but only as constituent parts of the DS3. Because the multiplexing of the DS1 signals into a DS3 is asynchronous and employs stuffing bits, the location of any DS1 in the SPE is not known. Therefore, an individual DS1 cannot be dropped and inserted without demultiplexing the DS3 in which it is carried. To overcome this limitation, lower speed asynchronous traffic, including DS1, is first assembled into synchronous VT blocks, which are then mapped into the STS-1 SPE.

13.7 VIRTUAL TRIBUTARIES

13.7.1 General

The various types of asynchronous traffic—DS1, DS1C, DS2, and CEPT E1—are placed into a VT format before being mapped into an SPE. Each of the asynchronous signals has its own VT format as shown in Table 13.2. In each case the VT rate has a considerable margin over the asynchronous traffic rate, and the traffic is mapped into the VT using stuffing bytes and other overhead information bytes. The VT bytes are arranged in nine rows, just like the arrangement of the SPE. It is the number of columns occupied by a VT that varies, but any VT can be mapped into a portion of the STS-1 SPE capacity. Because the location of any STS-1 in an OC-N stream can be identified, the location of any VT block in the SPE can also be found, and the VT with its asynchronous payload may be dropped and inserted at intermediate points along a SONET system.

Table 13.2
Virtual Tributaries

Asynchronous Signal		Virtual Tributary	
Type	Bit Rate (Mbps)	Type	Bit Rate (Mbps)
DS1	1.544	VT1.5	1.728
E1	2.048	VT2	2.304
DS1C	3.152	VT3	3.456
DS2	6.312	VT6	6.912

13.7.2 Virtual Tributary Formats

Each of the VT types has a frame duration of 125 microseconds, as shown in Figure 13.7. The VT bit rates shown are derived by multiplying the number of bytes in the particular VT by eight to obtain the bits, and multiplying by 8,000 frames per second. Since the VT frame has nine rows, the number of columns is equal to the total number of bytes in the VT divided by nine, and is three for VT1.5, 4 for VT2, 6 for VT3, and 12 for VT6.

13.7.3 Virtual Tributary Groups

A VT group consists of 12 columns, and seven VT groups form the payload of the STS-1 payload. Each individual group is made up of only one VT type; and four VT1.5 blocks, three VT2 blocks, or two VT3 blocks are byte-interleaved to fill one VT group. One VT6, of course, fills the VT group completely. The seven VT groups may be all of the same type, or they may include a mixture of types. Since there is very little DS1C or DS2 traffic in the networks and E1 is present only in international interconnections, the dominant VT type is the VT1.5 containing DS1 traffic. The formation of four types of VT groups is illustrated in Figure 13.8.

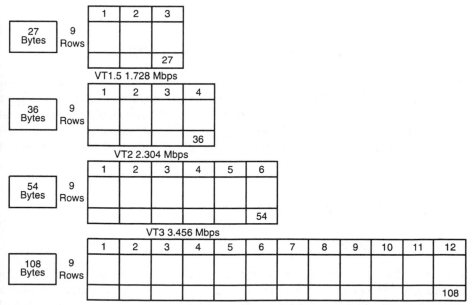

Figure 13.7 VT types.

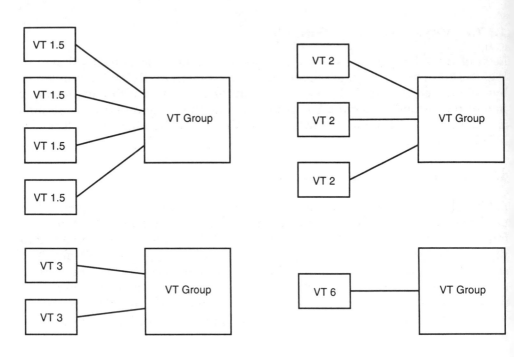

Figure 13.8 Formation of VT groups.

13.7.4 Virtual Tributary Group Mapping into SPE

Each VT group contains 12 columns of bytes, so the seven VT groups have a total of 84 columns. The STS-1 SPE has a total of 87 columns of available capacity, and the first column contains space for the POH bytes. Columns 30 and 59
are allocated to fixed stuffing bits, leaving the remaining capacity for the 84 VT
group columns. The structure of the resulting SPE is identical, no matter which
type of asynchronous traffic is mapped into the VT blocks.

13.8 SONET BROADBAND TRAFFIC

13.8.1 Concatenation

The traffic discussed up to this point has been asynchronous, and the type
with the highest speed is the DS3 with a bit rate of 44.736 Mbps, which is
mapped into one STS-1. The STS-N frame shown in Figure 13.4. has an SPE
capacity of 783 N bytes, which accommodates N SPE blocks, including the POH
blocks of each of the N STS-1 signals. These N SPE blocks remain together only

until the STS-N signal is demultiplexed into smaller capacity STS signals. In other words, the STS-N is channelized into STS-1 blocks, which do not necessarily have a common destination.

This form of multiplexing is not suitable for traffic requiring more capacity than one STS-1 can provide. If such traffic were broken down and mapped into multiple STS-1 streams, and then the STS-1 blocks were multiplexed up into a STS-N signal, this signal is quite likely to be demultiplexed along its path and the constituent STS-1 signals sent to different destinations. In order to handle this type of traffic, it must be mapped into a single SPE within one block. This block is called a concatenated signal, designated STS-Nc, and its optical version is an OC-Nc signal. The presence of the c shows that its payload is not channelized, and the SPE has to be delivered intact to its destination.

13.8.2 STS-Nc Frame Format

The STS-Nc frame format, shown in Figure 13.9, may be compared with the STS-N frame of Figure 13.4. In both cases, the transport overhead occupies 3 N columns, leaving 87 N columns for the SPE capacity. In the STS-N case, N columns of the capacity are needed for the N POH bytes corresponding to the N

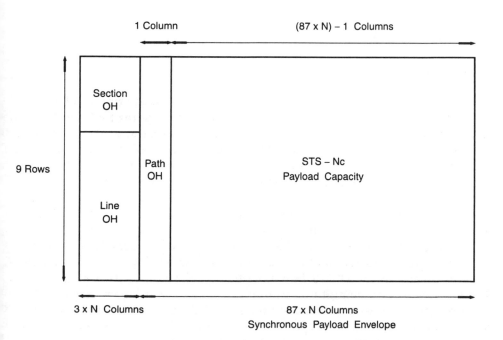

Figure 13.9 STS-Nc frame.

STS-1 signals in the payload, leaving 86 N columns as the payload capacity. In the STS-Nc format, only one column is needed for the POH bytes because the concatenated payload comprises one traffic block to be delivered intact over the network. The payload capacity is then $(87 N - 1)$ columns. To indicate that concatenated signals are being transported, some bits in the H1 and H2 bytes in the STS-Nc transport overhead carry a concatenation indication pattern.

13.9 OPERATIONS, ADMINISTRATION, MAINTENANCE, AND PROVISIONING

The operations, administration, maintenance, and provisioning (OAM&P) of SONET systems is much more sophisticated than the corresponding functions in the asynchronous systems in current use. The large capacity of the section, line, and path overheads makes system status and performance information available at local and remote sites and makes possible considerable reductions in operating costs, as well as improvements in operating efficiency.

This aspect of SONET systems is evolving steadily, and the standards that will define the future OAM&P provisions are in preparation. Chapter 11 includes details of the operational and control functions which will be included in the SONET network.

13.10 SONET SYNCHRONIZATION

13.10.1 General

SONET systems provide the means of synchronous transportation of payload traffic from one point on a network to another. This requires that all elements of the network employ timing sources that are traceable to a precision clock. The current asynchronous systems operate as point-to-point links, with asynchronous DS3 access points. Interconnection between links at intermediate sites in the network is at the individual DS3 level, involving DSX-3 frames. SONET systems must have synchronous interconnections so that a synchronous path across the network is maintained for whatever payload is being carried. The interconnection techniques are compared in Figure 13.10.

In the asynchronous case, the site employs back-to-back terminals, with all of the system capacity brought down to the DS3 level, and the interconnections are made at the DSX-3 frame. In the synchronous network, the OC-48 SONET systems are demultiplexed only to the OC-12 tributary stage, and the required cross connection is made optically. This represents a great saving in equipment, and because the DSX-3 frame and all of its coaxial cabling is eliminated, transmission quality and reliability are improved.

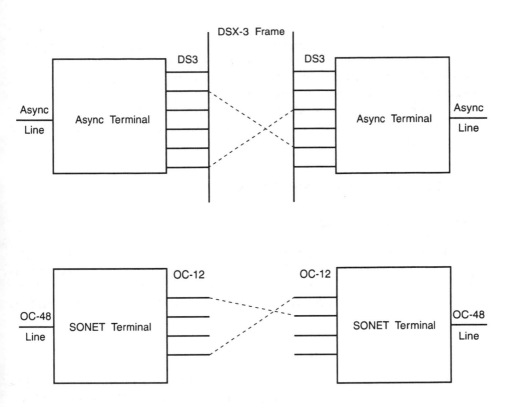

Figure 13.10 System interconnection.

13.10.2 Synchronization Methods

Synchronization is described in detail in Chapter 7. This section covers the techniques that may be used in SONET networks. There are five methods used for synchronization at various network elements: external, free running, line, loop, and through timing.

External timing is used at sites that are provided with a building integrated timing system (BITS). This is a source with a very high accuracy, and it is distributed within the site to all of the NEs. A method of BITS distribution that is commonly used is a duplicated DS1 signal, and all of the SONET bays have redundant input ports to accept the timing supply. Due to the high accuracy of the BITS timing signals, all of the sites on a network that employ this technique are synchronized to one another. This method has to be employed at any site where synchronous traffic originates or is cross-connected.

Free-running or internal timing uses a clock source in the NE to time the signals. It is not used where synchronous traffic originates or is cross-connected.

Line timing is used for an NE that is set up as an ADM. A clock is derived from the incoming line OC-N signal and is used to time all outgoing OC-N signals.

Loop timing is used only at small-capacity sites at the end of a link. The clock derived from the incoming OC-N signal is used to time the transmit OC-N signal sent back in the opposite direction.

Through timing is used at regenerator sites and for intra-office SONET links. The clock derived from the incoming OC-N is used to time the outgoing OC-N in the same direction of transmission.

13.11 SONET NETWORK PROTECTION

13.11.1 Equipment Protection

The methods of protecting a SONET system from equipment failure are similar to those used in asynchronous digital systems and in the DCSs. The DS1 interfaces are protected on a 1:*N* basis, and the DS3 and optical tributary OC-N interfaces are protected 1:1. Redundant pair protection is used for units such as timing distribution and power supplies.

13.11.2 Automatic Protection Switching

The purpose of automatic protection switching (APS) is to switch the traffic from a failed fiber-optic channel to a protection channel. The channel is defined as all of the fiber-optic line equipment from the point where the LOH is added to the point at the distant end where the LOH is terminated, including all of the regenerator and ADM equipment in the line.

Two basic forms of APS may be used on SONET systems: 1 + 1 and 1:*N*, where *N* has a minimum value of 1 and a maximum value of 14. Systems with one protection channel may be software configured for 1 + 1 or 1:1 operation. In the 1 + 1 configuration, both transmitters are connected online, and at the distant end the receiver selects one line as the working channel. If this line fails the receiver switches to the other line. This is similar to path diversity in a microwave radio system. The 1 + 1 protection is nonrevertive, which means that service does not revert back when the original line is restored. In the 1:*N* case, one protection channel is provided for *N* working channels. If any one working channel fails it is replaced by the protection channel, but any subsequent working channel failure is not protected. For this reason 1:*N* protection is revertive, and service returns to the original working channel when it is restored.

The APS function in a SONET system is controlled using messages conveyed in the K1 and K2 bytes of the LOH. The APS function provides service protection for equipment and line failure of a single working channel, but at the network level it cannot protect against the loss of multiple channels on a fiber link or the loss of a site. These types of problems may produce cata-strophic networkwide failures, unless more comprehensive restoration schemes are adopted. One very important concept in SONET is the use of SONET rings to provide network protection as described in the next chapter.

References

[1] Bellcore, Bell Communications Research, Technical Reference TR-TSY-000253, Synchronous Optical Network (SONET) Transport Systems: Common Generic Criteria.

[2] Bellcore, Technical Reference TR-TSY-000496, SONET Add-Drop Multiplox Equipment (SONET ADM) Generic Criteria.

[3] ANSI T1.105-1991, American National Standard for Telecommunications, Optical Interface Rates and Formats Specifications (SONET).

Survivability with SONET Rings

14

14.1 INTRODUCTION

In Chapter 12, the survivability of networks using the DCS 3/3 and the RDPS was described. This technique is dependent on the availability of DS3 interfaces at every site, which is a reasonable requirement where the network consists of asynchronous transmission systems. With the evolution of SONET networks, DS3 interfaces at intermediate sites on the network are no longer available, and the ADM provides the means for traffic to access the network at points between major sites. SONET systems can use their extensive overhead capacity to monitor the condition of the network, exchange information between network elements, and initiate the appropriate protection switching to restore traffic without human intervention.

There is increasing pressure for very rapid network restoral when a loss of traffic occurs. Much of the network traffic now consists of high-speed data services, where any interruption of appreciable length is a disaster for the customer—and therefore for the network operator who is likely to lose the customer. Guarantees of uninterrupted service are being sought and offered, and instead of the minutes involved in the DCS 3/3 and RDPS technique, restoral in fractions of a second is becoming the objective. Only a *self-healing network* approach, with no manual intervention, can satisfy this type of time-delay objective. The SONET-ring concept is being developed to allow this objective to be met, and it covers not only the loss of a transmission path but, in many cases, the loss of a network site.

14.2 SONET-RING CONFIGURATIONS

The three types of SONET-ring configurations—self-healing rings, unidirectional rings, and bidirectional rings—are described in the following sections.

14.2.1 Self-Healing Rings

Self-healing rings consist of a number of nodes interconnected by two-way transmission links to form a closed loop, as shown in Figure 14.1. The ring has sufficient transmission capacity or redundant equipment, or both, to allow traffic to be restored after a failure. SONET regenerators may be included in any of the links between the nodes.

Traffic sent from one node to another may travel in either direction around the loop. The direction of travel is either clockwise (CW) or counter clockwise (CCW).

14.2.2 Unidirectional Rings

In a unidirectional ring under normal working conditions, every node on the ring transmits toward another node in the same direction, either CW or CCW. For example, in Figure 14.1, traffic from node A to node D may be routed CCW around the left-hand side of the ring, and traffic from node D to node A will then also travel CCW around the right-hand side of the ring. The duplex traffic between any two nodes traverses the entire length of the ring when the unidirectional method of operation is used.

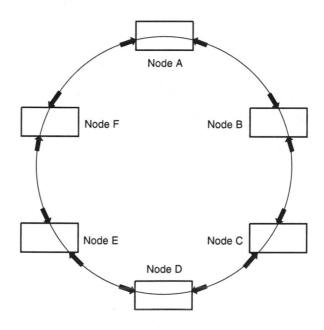

Figure 14.1 Self-healing SONET ring.

14.2.3 Bidirectional Rings

A bidirectional ring, under normal working conditions, carries each side of a duplex connection in opposite directions around the ring. Referring again to Figure 14.1, traffic between nodes A and E may occupy both directions of the portion of the ring between the two nodes on the left-hand side of the ring. The routing chosen for the traffic between two nodes is normally over the shorter path, and in contrast to the unidirectional ring the traffic does not traverse the entire ring.

14.3 RING-PROTECTION METHODS

14.3.1. Introduction

Protection of the rings is based on the extensive information about system status and performance available in the overheads of the SONET links. Two types of protection techniques have been evolved: line switching and path switching. In the line switching case, the line extends only between two adjacent nodes, and the overhead is monitored at each of the nodes. Abnormal conditions cause the switchover of service from the working to the protection SONET system line. Where path switching is used, the path overhead conditions of each STS-1 facility are monitored separately. Each STS-1 is carried over two ring paths, and one of the signals is selected based on the path-overhead indications. If the selected path deteriorates, service is transferred to the other path. The switching is carried out independently on each STS-1 facility.

14.3.2 Line Switching

Line-switched rings, which may be unidirectional or bidirectional, use the line-overhead indications to monitor the system performance. The detection of a line failure on the ring initiates the exchange of messages among the nodes. The end result of these exchanges is the initiation of line-protection switching at the appropriate nodes to enable traffic to be restored. Line switching is not initiated by any indications of path performance degradation.

Line-switched rings, whether unidirectional or bidirectional, are limited to a maximum of 16 nodes.

14.3.3 Path Switching

In contrast to line switching, two unidirectional rings are used for the path-switched configuration, with each ring acting as protection for the other. The 1 + 1 protection switching technique is used, and the STS-1 level traffic is bridged on to both of the rings at the originating node, with one signal traveling CW and the other, CCW. Signals from both rings are present at the terminating node, and one of these STS-1 signals is selected as the working signal based on the POH indications. Failure of this path causes automatic switchover to the received STS-1 signal from the other ring. Each STS-1 is protected separately, and may travel on either ring.

14.4 UNIDIRECTIONAL PATH-SWITCHED RINGS

14.4.1 Normal Operation

A unidirectional path-switched ring (UPSR) arrangement is shown in Figure 14.2. The nodes, each of which has an ADM, are interconnected using two fibers to form two counter-rotating unidirectional closed rings, and the line signal speed can be any of the available OC-N rates. As mentioned in Chapter 13 on SONET, an OC-N system has a transmission capacity of one STS-N or N STS-1 facilities.

Figure 14.2 illustrates the routing of an STS-1 signal, under normal conditions, between nodes B and E. At node B, the outgoing STS-1 signal is routed in the ADM to the line system on fiber 1 in the CW direction and to the line system on fiber 2 in the CCW direction. At node E, the STS-1 signal is received from the line systems on both fibers, and the ADM selects one of these as the working STS-1 signal. In this illustration, it is assumed that the signal from fiber 1 is selected.

Other STS-1 paths can be established between the nodes on the ring, and each of these may select the line system on fiber 1 or fiber 2 as its working path. Once an STS-1 path is established, it occupies that time slot of the OC-N capacity on every link of both fibers. The total number of STS-1 links that can be established on the ring between any nodes is limited to the value of N.

14.4.2 Individual STS-1 Failure

The performance of each STS-1 is monitored in its POH, and if deterioration beyond acceptable limits is detected, the service is switched to the alternate signal on the other fiber. Each STS-1 is routed independently of the others, and a route change on one has no effect on the other STS-1 traffic on the ring.

14.4.3 Fiber-Optic Cable Break

In the event of a cable break that affects both fibers in a link, both unidirectional rings are broken and the path overheads of all the STS-1 signals routed over the affected path indicate loss of service. In the case shown in Figure 14.3,

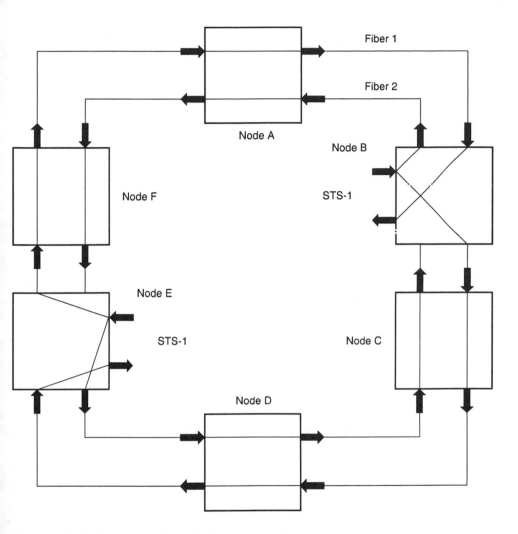

Figure 14.2 Unidirectional path-switched ring – normal.

a cable break between node C and node D removes the transmission path between nodes B and E via the right-hand part of the ring. All of the STS-1 traffic between nodes B and E now travels on the duplex path on fibers 1 and 2 along the left-hand part of the ring. The traffic is unprotected until the cable break is repaired, so any other equipment or fiber-cable failure on the ring that occurs during this time will cause loss of service.

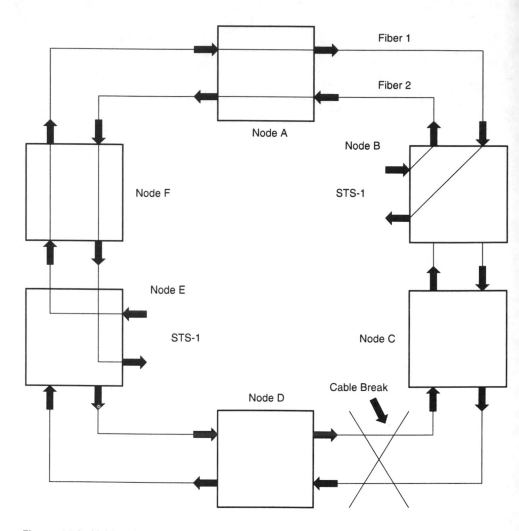

Figure 14.3 Unidirectional path-switched ring – cable break.

14.4.4 Restoration Time

No communication is needed between the nodes to initiate switching to the protection routing to restore the traffic, and no knowledge of the ring configuration is needed at any of the nodes. For these reasons the time to complete traffic restoration is less than 50 ms for the path-protected unidirectional ring.

14.5 UNIDIRECTIONAL LINE-SWITCHED RINGS

14.5.1 Normal Operation

A unidirectional line-switched ring (ULSR) is shown in Figure 14.4. The nodes are interconnected in the same manner as the UPSR arrangement described in Section 14.4, but in this case—in normal operation—one unidirectional ring carries the working traffic, and the other is reserved for protection of the working ring.

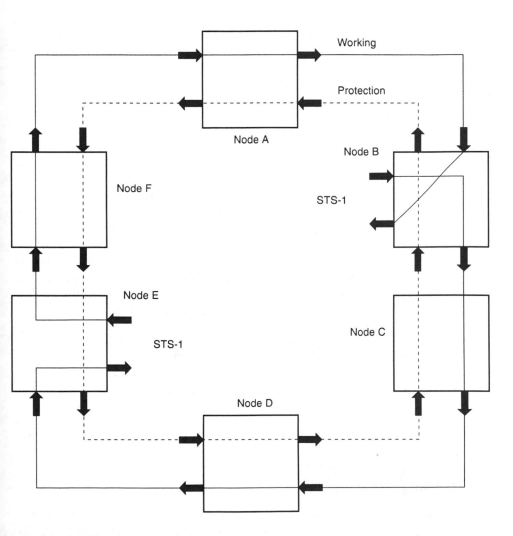

Figure 14.4 Unidirectional line-switched ring – normal.

The path of an STS-1 between nodes B and E under normal working conditions is illustrated, and for this example it is carried on the working fiber in the CW direction. It traverses the entire length of the ring, so this time slot cannot be reused, and the traffic capacity of the ring is limited to N, where an OC-N SONET system is used as the transmission link.

14.5.2 Fiber-Optic Cable Break

In the event of a cable break affecting both fibers, both unidirectional rings are broken. All of the traffic is interrupted because the traffic was traversing the whole length of the working fiber. The failure is indicated in the LOH of each of the links, and the restoration process is initiated. The process makes use of the line automatic protection system (APS), which is described in detail later in this chapter.

The indications from the LOHs cause messages to be exchanged among the nodes, and the necessary protection switching actions are carried out to restore all of the traffic on the ring. Figure 14.5 shows a cable break between node C and node D, affecting both fibers. The protection switching loops the fibers back at these nodes, which are on each side of the break. A complete loop is formed and all of the nodes have access to it, so all of the traffic is restored but is now being carried on an unprotected path.

In the event of another cable break, the APS will cause loopback of the fibers on each side of that break. This will divide the ring into two segments, and the amount of traffic will be greatly reduced because communications will be lost between the segments.

When the cable break is repaired, all of the links between nodes resume normal transmission, which results in the LOHs indicating proper system performance. When these conditions are recognized at the nodes, the loop switching is removed and the ring reverts to its normal operating conditions.

14.6 BIDIRECTIONAL LINE-SWITCHED RINGS

14.6.1 Types of Bidirectional Line-Switched Rings

There are two types of bidirectional line-switched rings: two-fiber and four-fiber. Each of these is described in later sections, but both share some common characteristics.

14.6.2 Traffic Routing

In a bidirectional ring, the traffic between any pair of nodes can be routed CCW or CW around the ring, since both directions of transmission travel over the same path. It is normal to route traffic over the shorter of the two available

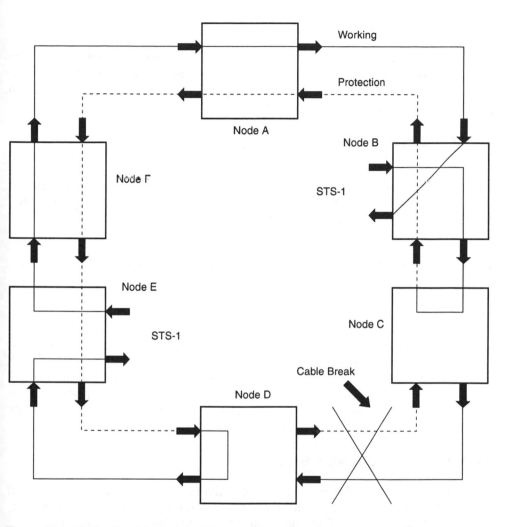

Figure 14.5 Unidirectional line-switched ring – cable break.

paths. Each STS-1 between two nodes occupies a transmission time slot only in the spans it traverses. It does not travel the entire length of the ring, as it would in a unidirectional ring. The total number of STS-1 paths that can be established between various pairs of nodes is limited by the OC-N capacity of the spans, but it may exceed the number that can be loaded on a unidirectional ring in which every STS-1 traverses the entire length of the ring.

Although the shorter path between a pair of nodes is normally selected as the traffic route, it may be advantageous to use the longer path in some cases. This is because a particular span may already be carrying its maximum load but capacity is available on each of the spans of the alternative route.

14.6.3 Extra Traffic

When the protection capacity of the ring is not being used for restoration of the working traffic, it may be used to carry extra traffic. Extra traffic has a lower service priority because it is preempted by the normal working traffic when protection switching occurs.

14.6.4 Restoration Time

For rings that have a total length of less than 1,200 km and are not carrying extra traffic, the objective for restoration time is 50 ms or less. Additional time is needed for cases that do not meet these requirements, and this may cause the total interval to extend beyond 100 ms.

14.7 TWO-FIBER BIDIRECTIONAL LINE-SWITCHED RINGS

14.7.1 Normal Operation

In a two-fiber bidirectional line-switched ring (BLSR), each of the fibers in the spans carries both working and protection channels in the same direction, as illustrated in Figure 14.6. Of the N STS-1 channels in the OC-N system on each fiber, half are assigned as working capacity and half as protection capacity.

The path of an STS-1 between nodes B and E is shown under normal conditions. It is carried on one of the STS-1 slots assigned to the working capacity of the two fibers on each of the spans between these two nodes.

14.7.2 Fiber-Optic Cable Break

In the event of a cable break affecting both fibers in the path between nodes B and E, the traffic between these two sites is interrupted. The failure is indicated in the LOHs on all of the spans, and the restoration process is initiated.

The indications from the LOHs cause messages to be exchanged among the nodes, and the necessary switching actions are carried out to restore all of the traffic on the ring. Figure 14.7 illustrates a cable break between node C and node D, which affects both fibers. Switching takes place at the nodes on each side of the break. In each case the working channels are looped back to the protection channels on the fiber in the opposite direction of transmission.

This action causes a complete loop to be formed, providing access to all sites. All of the traffic is then restored. However, because the protection capacity of the ring is now in use, the ring is operating on an unprotected basis.

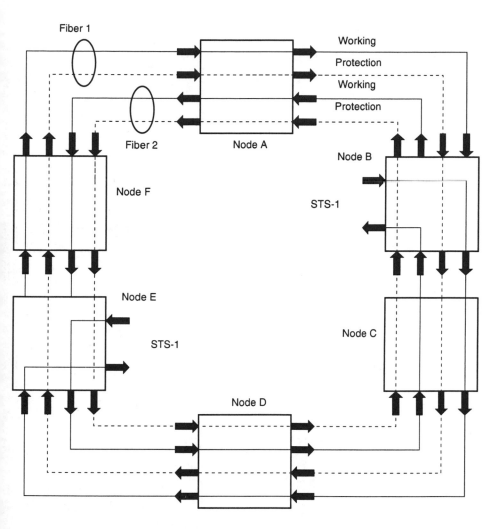

Figure 14.6 Two-fiber bidirectional line-switched ring – normal

When the cable break is repaired, all of the spans return to normal, which is indicated in the LOHs. When these conditions are recognized at the nodes, the loop switching is removed and the ring returns to its normal operating state.

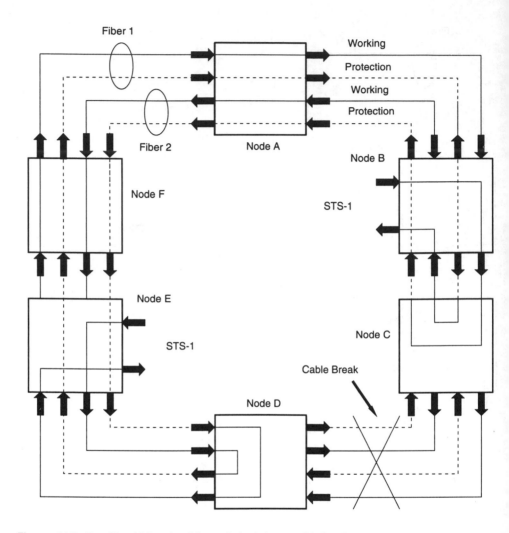

Figure 14.7 Two-fiber bidirectional line-switched ring – cable break.

14.8 FOUR-FIBER BIDIRECTIONAL LINE-SWITCHED RINGS

14.8.1 Normal Operation

In a four-fiber bidirectional line-switched ring (BLSR), two of the fibers carry the working channels on a duplex basis, and the other two carry the protection channels, as shown in Figure 14.8.

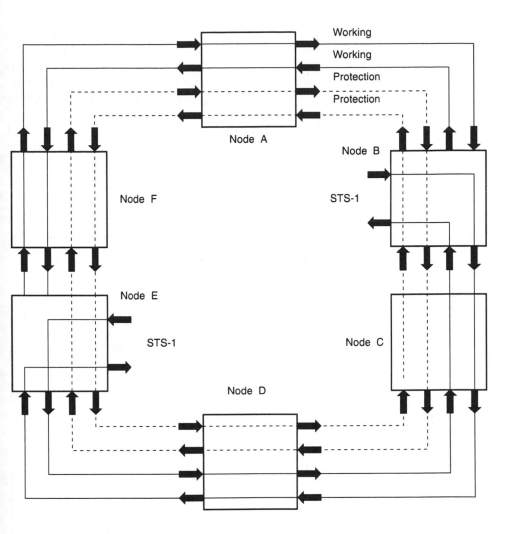

Figure 14.8 Four-fiber bidirectional line-switched ring – normal.

The path of an STS-1 between nodes B and E is shown under normal conditions. It is carried on one of the STS-1 slots on the working fiber system over each span in its path.

14.8.2 Fiber-Optic Cable Break

In the event of a cable break affecting the working and protection fibers between nodes B and E, the traffic between these two sites is interrupted. The failure is indicated in the LOHs of all of the spans, and the restoration process is initiated. The indications cause messages to be exchanged among the nodes, and the necessary switching actions are carried out to restore all of the traffic on the ring. Figure 14.9 shows a cable break between node C and node D affecting all four fibers. Switching takes place at the nodes on each side of the break. Both directions of the working system are looped onto the protection fibers in the opposite direction of transmission.

This action causes a complete loop to be formed, providing access to all sites, so all the traffic is restored. Because the protection fibers are in use, the ring is now operating on an unprotected basis.

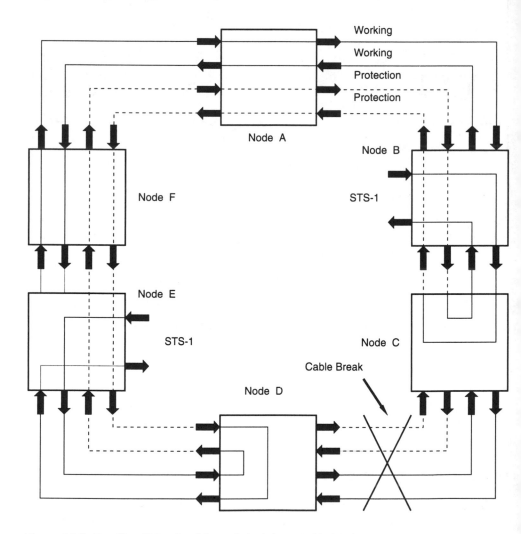

Figure 14.9 Four-fiber bidirectional line-switched ring – cable break.

All of the spans return to normal when the cable break is repaired, which is indicated in the LOHs. When this condition is recognized at the nodes, the switching is removed and the ring returns to its normal operating state.

14.8.3 Span Failure

In the discussion of fiber-cable breaks in the various types of rings, it has been assumed that a complete cut has occurred, involving all of the fibers. The restoration of the traffic has made use of what is known as a ring switch.

In the case of the four-fiber line-switched ring, the failure of the working system in a span is capable of being handled with a span switch. A span is defined as the transmission system between two nodes on the ring. The failure may be due to an electronic problem or a cable break affecting only the fibers carrying the working channels. Whatever the cause, when the failure is indicated in the LOHs, the traffic on the working path is switched to the protection path at both ends of the span. This does not affect any other span, and span switching may occur independently on any of the spans on the ring. A complete cable break on any span invokes ring switching, which overrides any span switches already in place.

14.9 RING AUTOMATIC PROTECTION SWITCHING

14.9.1 General

The standards for ring automatic protection switching (APS) are still evolving, and the fine details are subject to modification. The description given here covers the basis of the APS functions, which are unlikely to change to any extent.

14.9.2 SONET Line Overheads

The APS makes use of the LOH bytes K1 and K2 that are defined in the first STS-1 block of the OC-N. Where a four-fiber ring is used, the APS feature is active only on the protection fibers. The functions that are required on a real-time basis to affect protection switching are carried in the K1 and K2 bytes. Other functions necessary to support the APS are carried in the line data communications channel (DCC).

14.9.3 Node Designations

Each node on a ring is assigned an identification number, ranging from zero to 15, and this number is independent of the order in which the nodes are situated around the ring. All rings have a maximum of 16 nodes.

Each node on the ring stores a map of the ring configuration, including the details of the routing of every STS-1 from node to node. This information is updated by personnel on site or through the OSS from the control center. These maps must be maintained with the current ring status to avoid misconnections when traffic is being restored.

14.9.4 Normal Conditions

When a ring is operating normally with no fiber-optic cable or equipment problems, each node indicates on the K bytes in each direction around the ring that it has no switch request and includes its own node identification number.

14.9.5 Initiation of Switching

Switching is initiated automatically by a signal failure or a performance-degradation condition on the ring. When a node experiences indications that a switch is necessary, it sends a *bridge request* message in the K bytes in both directions around the ring over the short path and the long path. The message includes the identification number of the node originating the message, known as the source node, and the identification number of the node to which it is being sent, known as the destination node. The destination node is the node adjacent to the source node, on the other side of the failed span. Nodes on the ring, other than the destination node, receive the bridge request and enter a pass-through state, which allows the K bytes containing the message to continue on their way to the destination node without modification.

When the bridge request is received at the destination node, the working channels that would have continued over the failed span are bridged onto the protection channels in the opposite direction of transmission. This is illustrated for the four-fiber bidirectional ring in Figure 14.9.

14.9.6 Restoral to Normal

When the failed span has been repaired, the source nodes stop sending the bridge request messages round the ring. The nodes other than the destination node stop passing through the K bytes, the bridging is removed at the destination node, and the traffic reverts from the protection channels to the working channels, thus restoring the ring to normal operation.

14.10 ACCESS TRAFFIC SURVIVABILITY

The ring configurations provide traffic survivability in the event of transmission failures between the nodes. However, the loss of the link between the source of the access traffic and its associated node results in all of the outgoing and incoming traffic of that node being disconnected from the ring.

Dual feeding, where the traffic from one source is handled by two adjacent nodes, can be used to provide access traffic survivability. The technique is shown in Figure 14.10, where the STS-1 block handled by node B is also connected to node C. The traffic is sent into the ring at both of these nodes, and a switch at node C selects one of the two signals for transmission around the ring to node E. The STS-1 block sent into node E is dropped at site C and continues on to node B. This configuration is called *drop and continue*. The loss of the

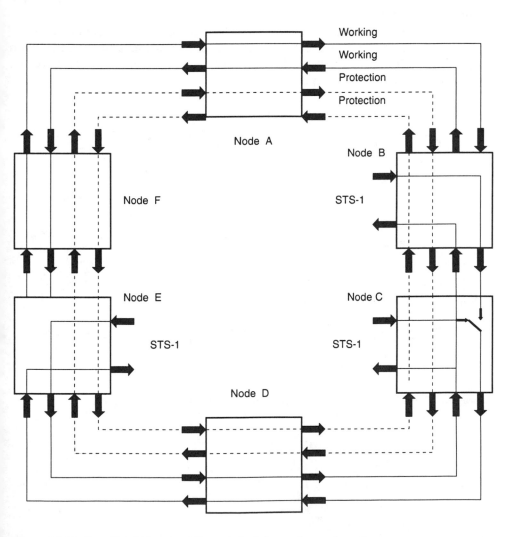

Figure 14.10 Four-fiber bidirectional line-switched ring – drop and continue.

access traffic link to either node C or node B does not cause the traffic to be lost because the link to the other node is still intact.

14.11 INTERLOCKING RINGS

Rings in a network do not normally exist in isolation, and a node may be common to more than one ring, as illustrated in Figure 14.11(a). The SONET ADM equipment in each of the rings at the common node is interconnected at either the optical or the STS-1 level, allowing paths to be established between nodes on both rings. This configuration has a major weakness because the loss of the common node would break both rings.

This weakness can be overcome by interconnecting the rings at two adjacent sites, as illustrated in Figure 14.11(b). If either of these nodes fails, each ring will go into the appropriate ring-protection mode, and the links between the rings will remain operational.

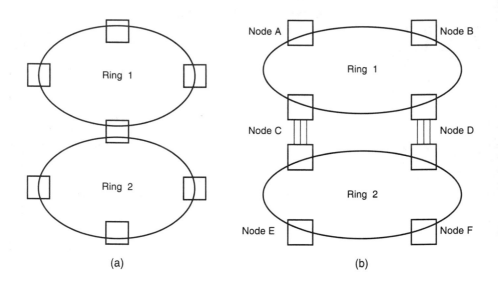

(a) (b)

Figure 14.11 Interlocking rings: (a) one common node, and (b) two common nodes.

Future Network Services and Technology 15

15.1 GENERAL

Advances in telecommunications technology and the development of new services have resulted in fundamental changes in network topology in the years since 1980. Assuming that this rapid pace of change is maintained, it would be foolhardy to attempt to forecast the network configurations and all of the services that may be provided by the year 2010.

It is possible, however, to examine the trends in network technology and the likely demands for broadband services to be introduced over the next few years, and from these to deduce the directions that the next stage of evolution of long-distance telecommunications networks may take.

In general, the networks of the future may be expected to employ synchronous transmission links, be controlled and monitored from central locations, and carry integrated broadband services on a *bandwidth on demand* basis. These topics are discussed in more detail in the rest of this chapter.

15.2 CHARACTERISTICS OF CURRENT NETWORKS

Current digital networks carry narrowband services at rates of 1.544 Mbps or less, with interfaces at the DS0, DS1, and DS3 levels. Voice and low-speed data make up a large but slowly declining part of the total network load. Traffic is carried in 8-bit bytes between two points on the network, and in most cases the circuit is set up at the beginning of the call, using the network signaling system, and is broken down again at the end of the call. The exception is the leased line, where the circuit is set up when the lease is initiated and is available to the customer for traffic for the duration of the lease. These leased lines may be routed and connected using the remotely controlled DCS 1/0 systems in a network.

Switched traffic is handled by toll switches, which are provided with built-in routing tables used to determine the path of each individual call through the network. Special traffic, such as video-conference calls, is set up on an as-needed basis, and may also use the DCS 1/0 switches for connection and disconnection purposes. In each case, the circuit that is established is dedicated full-time either to the call until it is completed or to a customer for the duration of the lease. This piece of network capacity is tied up for a certain period and is not available for any other traffic, even if it is actually utilized to carry information for only a part of the time.

15.3 DEPLOYMENT OF SONET SYSTEMS

The deployment of the SONET systems described in Chapter 13 is starting to take place on a link-by-link basis, but the complete replacement of the asynchronous digital systems may be expected to take a few years [1]. In the meantime, both asynchronous and synchronous systems offer DS1 and DS3 interfaces, which will continue to handle the current types of narrowband traffic. Completely synchronous paths between sites will not be available until the sites are linked by SONET transmission systems, but then these paths may be used to carry a variety of broadband services using transmission speeds in excess of the primary rate of 1.544 Mbps.

The SONET transmission systems provide optical interfaces at the OC-3 and OC-12 levels within the sites, and other network elements such as cross-connect switches are being introduced with optical interfaces in place of the metallic DS1 and DS3 ports. The intrasite connections between network elements will use fiber-optic cables in place of the balanced pair and coaxial links in current use, thus reducing the bulk and weight of the cables and making induced noise and crosstalk much less of a problem. The access traffic between local and long-distance carriers will also be carried on SONET systems and will be groomed using DCS 3/1 cross-connect systems.

15.4 POTENTIAL BROADBAND SERVICES

15.4.1 Types of Services

Three applications for broadband services are already identified:

- Interconnection of local area networks (LAN);
- Image transfer;
- Video transmission.

15.4.2 Local Area Networks

LAN techniques were developed to permit computer terminals and other data terminals to be interconnected within an office, a building, or a campus containing a number of buildings. Metallic cables, used to interconnect the terminals, commonly use a transmission rate of 10 Mbps within the LAN.

The interconnection of geographically remote LANs requires the use of the long-distance network, which, as we have seen, currently provides only narrowband transmission facilities at rates up to 1.544 Mbps. The LAN traffic can be *throttled down* to pass over the narrowband facilities, but it is then subject to delays, which may make interactive operation impracticable.

15.4.3 Image Transfer

The transfer of images, such as medical X-rays and graphics, over the network is not feasible using the present narrowband facilities. Transport of film by vehicle is the only method available. Scanning the images electronically generates signals at rates of many megabits per second and attempting to compress the bandwidth to fit into the narrowband channels introduces intolerable transmission delays.

15.4.4 Video Transmission

Low-definition video services, such as video phone and video conferencing, employ bandwidth compression techniques that allow their transmission over the present narrowband network facilities. Full-motion broadcast television to the NTSC standards and the high-definition television (HDTV) service projected to be introduced in a few years will require transmission rates higher than the narrowband facilities can handle.

15.4.5 Broadband Service Introduction

None of these potential services can be carried over the current long-distance network, and their successful introduction must await the availability of broadband network service capability to be provided by the broadband integrated services digital network (ISDN).

15.5 BROADBAND ISDN

The first type of ISDN to be introduced was a narrowband version designed to carry voice and data traffic over the current network. The broadband integrated services digital network (B-ISDN) concept is being evolved to provide standard user-to-network interfaces, transmission rates, switching, multiplexing, and signaling for integrated voice, data, video, and imaging services.

A major requirement is that the switching, multiplexing, and signaling equipment used in the B-ISDN has to be flexible so that it can evolve to accommodate new future services without requiring any equipment replacement. B-ISDN is being designed to use the SONET transmission systems currently being deployed. The speeds to be used for the transmission of B-ISDN are DS3 at 44.736 Mbps, STS-1 at 51.84 Mbps, STS-3 at 155.52 Mbps, and STS-12 at 622.08 Mbps. These last two SONET rates are common with the STM-1 and STM-4 levels of the European SDH synchronous systems, which will facilitate international interconnection of B-ISDN traffic.

The technique that has been selected to implement the B-ISDN is asynchronous transfer mode (ATM), and this is outlined in the next section.

15.6   ASYNCHRONOUS TRANSFER MODE

15.6.1   ATM Basics

The ATM technique integrates the various types of traffic services into a common format and transfers them across the network in 53-byte cells, which have a 5-byte header and 48 information bytes. It employs new multiplexers, switches, and cross-connect systems, which, when integration is complete, will replace all of the existing multiplexing, switching, and cross-connect equipment in the current network. ATM will be deployed initially as an overlay network to handle broadband services such as imaging, and the narrowband voice and data traffic will continue to use the existing techniques for many years, due to the very large investment in the existing plant. The SONET transmission systems will carry both the existing and ATM-format traffic and will not need to be replaced as additional services are provided using ATM.

The existing network sets up end-to-end paths dedicated to a user for the duration of a call. The path has a fixed bit rate or bandwidth and cannot be used for other traffic even if the user's information is not using the bandwidth continuously. With ATM, a number of users occupy a bit stream in turn, and each cell includes a *header*, which contains routing information so a dedicated path is not required. The ATM format allows a *bandwidth-on-demand* approach to be used to accommodate traffic using various different bit rates.

15.6.2   The ATM Cell

The ATM cell consists of a header that is five bytes long, followed by 48 bytes of information capacity, as illustrated in Figure 15.1. The information section can carry voice, data, image, video, or new services yet to be defined. The cell

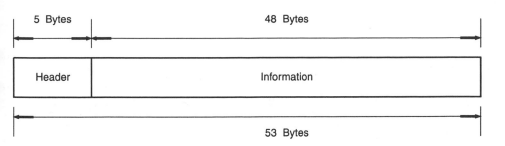

Figure 15.1 ATM cell structure.

is of fixed length for all of these types of service, in contrast to the variable-length cells used in the frame relay technique, which is best suited to bursty data traffic.

The use of a longer cell would be advantageous for high-speed data transfer from one computer to another, since the overhead content of the transmission would be less. However, to minimize delay in voice and video services, a shorter cell is necessary. The 53-byte ATM cell is a compromise between these two conflicting requirements.

15.6.3 Asynchronous Aspect of ATM

An ATM multiplexer generates a continuous stream of cells. Some of these cells contain traffic information, and the others are empty. Traffic generated by a particular user and sent into the multiplexer is assigned to cells on an as-needed basis, and there is no synchronization between a user's information and the positions the corresponding cells occupy in the continuous stream. It is the header bytes, rather than the position of the cell in the bit stream, that contain the cell routing information.

15.6.4 ATM Multiplexing

The ATM multiplexer shown in Figure 15.2 accepts a variety of digital traffic and produces a continuous stream of cells, each of which contains 53 bytes. The number of cells associated with each traffic source depends on the bit rate of the input traffic. The continuous-output stream will operate at one of the rates listed in Section 15.5, but the STS-3 level at 155.52 Mbps is most likely to be the initial transmission system interface. Therefore, it is used here to illustrate the multiplexing process.

The SONET format used for the ATM application is the STS-3c concatenated signal defined in Chapter 13, which has a payload capacity of 2,340 bytes per 125-microsecond frame. The concatenated format is used because the payload cannot be channelized into STS-1 blocks but has to be maintained

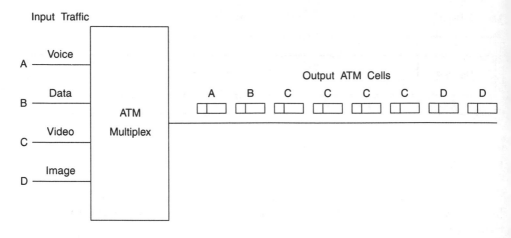

Figure 15.2 ATM multiplexing.

intact from source to destination. The ATM cells are mapped into the payload byte positions of the STS-3c frame. Because each ATM cell has 53 bytes, the STS-3c signal can hold a little more than 44 cells per frame. This means that a cell may be divided between the last few bytes of one frame and the first few bytes of the next frame. The information capacity of each frame is 44 cells, each containing 48 bytes or $48 \times 8 = 384$ bits, so each frame has $384 \times 44 = 16,896$ bits. The corresponding bit rate is 16,896 bits in 125 microseconds, or approximately 135 Mbps. The 5-byte cell headers require a total of 44 cells $\times 5 \times 8 = 1,760$ bits, which in a 125 microsecond frame require a bit rate of approximately 14 Mbps.

The total information in the service traffic that can be handled by the ATM multiplexer is limited to 135 Mbps when the output of the multiplexer is in the STS-3c format. The input traffic might consist of one broadcast-quality video source with a bit rate of 135 Mbps, or 44 video-conference inputs each at a rate of 3.1 Mbps, or any combination of input traffic that does not exceed 135 Mbps in total. Each input is allocated the appropriate number of cells in each frame, and the multiplexing process is flexible to allow the introduction of new services. A 600-Kbps data service occupies only one cell every five frames, for example, so ATM is not limited to handling only broadband service offerings.

Bursty data traffic is organized in packets that are assembled only when data is ready to be sent, and they do not necessarily have a fixed rate of transmission. These packets are handled in ATM by allocating the packet data to cells that are not in use for other input traffic, until all of the packet has been transmitted.

15.6.5 Virtual Circuits

Two types of virtual circuit (VC) are used with ATM. A permanent virtual circuit (PVC) is established using *nailed-up* connections between two ATM multiplexers. It is left in place for a long period and is used for the private-line type of services. A switched virtual circuit (SVC) is established on demand and exists for the duration of the call.

Each ATM cell header includes routing information called the virtual path identifier (VPI) and the virtual channel identifier (VCI), which show the VC to which the cell belongs. The VPI/VCI identifies an individual VC on a transmission link in the same way that the DS3 and DS1 numbers identify a facility assignment on an asynchronous digital link. The VPI/VCI values are reassigned when an ATM switch or cross-connect is traversed. An end-to-end VC is carried on a series of transmission system facilities, and it is identified by the unique VPI/VCI value allocated to it on each of the facilities.

15.6.6 Connectionless Service

The VCs described in the previous section are called connection-oriented, because both permanent and switched circuit connections are set up over a particular route when the circuit is initiated. A connectionless communication is an isolated event, such as the transfer of a block of data, and has no relation to any other traffic that may be sent between the same two points on the network.

The connectionless ATM cell has a header that denotes the cell as part of a connectionless communication and has its routing information and its source and destination addresses included in its information bytes.

15.6.7 ATM Cross Connects

ATM cross connects are for use in PVC applications, and will be used initially to provide only virtual path cross connections. Further evolution may allow VC cross connections to be handled.

15.6.8 ATM Switches

The function of ATM switches is to switch ATM cells, using the information in the cell header. Initially they will support only some of the B-ISDN services and provide an STS-3 rate SONET interface, but STS-12 SONET and DS3 rate interfaces may be included in the future. The services likely to be supported in initial applications include:

- PVCs for the interconnection of private LANs;
- SVCs to support video and imaging services;
- Packet data services.

Since the standards for ATM switching and signaling are not yet complete, it is likely that vendor proprietary schemes will be incorporated in the earlier switches deployed in the network.

15.7 NETWORK EVOLUTION WITH ATM

The standards for ATM are still evolving, and the protocols for the cell-header and information bytes, as well as the operational aspects of the use of the ATM technique for the provision of B-ISDN services, remain to be resolved. ATM switches have been produced for field trials, which will aid in refining the ATM techniques.

It is likely that the introduction of ATM equipment will form an overlay network superimposed on the existing structure, but the SONET transmission systems will be common to both ATM and the existing network elements. The concentration of development in the ATM area will be on the provision of those B-ISDN services that cannot be supported by the present network. Eventually the narrowband traffic will be transferred to ATM, but, due to the investment in existing plant, the voice traffic is unlikely to migrate to ATM for many years. It may well be that the transition will not be complete in the next decade. When the transition to ATM is finally complete, the only elements of the current network to survive will be the SONET transmission systems.

References

[1] Bellcore, Technical Reference TR-TSY-000253, Synchronous Optical Network (SONET) Transport Systems: Common Generic Criteria.

About the Authors

William S. Lee was born in China and grew up in the Philippines. He received a B.Sc. degree from the Polytechnic Colleges of the Philippines, an M.Sc. from the University of California, Berkeley, and a Ph.D. from Stanford University, all in electrical engineering. He has been with several communications companies in various capacities, including 15 years with GTE Lenkurt, mostly in their Advanced Development Laboratories. In 1985 he joined GTE Sprint, which evolved into the long-distance portion of Sprint Corporation, and is now a senior member of the technical staff in their Advanced Technology Laboratories.

Derrick C. Brown was born and raised in England, and received a B.Sc. and an M.Sc., both in engineering, from the University of London. He joined GTE Lenkurt in California in 1961 but returned to England in 1967, working for Plessey Telecommunications Research until 1973. Back again in California, he was with Farinon Electric until he moved to Southern Pacific Communications in 1979. That operation finally evolved into the long-distance portion of the Sprint Corporation, and he was in their Advanced Technology Laboratories until he retired in 1993.

Glossary

Add/drop multiplexer (ADM) – a network element that allows part of a high-speed digital stream to be dropped or added at an intermediate point in a transmission link, without demultiplexing and remultiplexing the entire stream.

American National Standards Institute (ANSI) – an organization that establishes standards for use in a wide range of scientific, medical, industrial, and technical applications, including the telecommunications industry.

Automatic protection switching (APS) – a system whose function is to switch the traffic from a working transmission path to a protection path when the working path fails or its performance deteriorates beyond allowable limits.

Asynchronous transfer mode (ATM) – a technique used to multiplex traffic of various speeds into fixed-length cells.

ATM Cell – a block of digits with a fixed length of 53 bytes, of which five are the header and 48 contain traffic information.

Broadband integrated services network (B-ISDN) – a network that is evolving to allow integrated voice, data, video, and imaging services to be carried on an integrated basis on long-distance networks.

Bit error rate (BER) – the probability of an error occurring per bit transmitted. It is a measure of performance of a digital link or transmission system.

Binary numbers – a numbering system consisting of ones and zeros. It is used extensively in computer processing and digital communications systems.

Bipolar signal – a digital signal technique that uses alternating positive and negative pulses for one state and zero value for the other state.

Bit – contraction of binary digit. In data transmission, a bit is the smallest unit of information and can be one of the two binary characters – one or zero – that computers use to perform calculations.

Bit rate – speed at which digital signals are transmitted, usually expressed in bits per second (bps) or multiples such as kilobits per second (Kbps) or megabits per second (Mbps).

Broadband – communications channel or service having a bandwidth or bit rate greater than a certain value. For digital channels the limit may be set at the DS1 rate or higher, but another common usage sets it at the DS3 rate or higher.

Buffer – a device used for the temporary storage of digital data bits. A buffer allows network elements having small differences in clock speed to operate together.

C-bit parity – a version of the DS3 signal, in which the C-bits may be used for in-service performance monitoring and other administrative functions.

Consultative Committee for International Telegraph and Telephone (CCITT) – an international organization that develops telecommunications standards.

Channel bank – area where the first stage of multiplexing is performed. In the transmit direction, multiple VF channels are modulated into a common higher frequency band or higher speed signal. In the receive direction, the equipment separates the signal back into individual channels. The analog version handles 12 VF channels and the digital type, called the PCM channel bank, multiplexes 24 channels.

Clock holdover – the capability of a timing clock to maintain frequency accuracy after all external timing references are lost.

Clock strata – the clocks used in a digital network are ranked by stratum number from 1 to 4, in descending order of stability.

Central office (CO) – a telephone office where subscriber lines are terminated and where the switching equipment for those lines is located.

Concatenation – *see* synchronous transport signal level *N* concatenated (STS-Nc)

Central processing unit (CPU) –core or central unit of a computer system; the input/output logic device that controls and accomplishes the arithmetic manipulations.

Cyclic redundancy check (CRC) – an error-checking technique in which the recipient of a digital signal frame calculates a remainder (by dividing the frame contents by a prime binary divisor) and compares this calculated remainder (called the CRC) with the value sent by the originating node.

Digital cross-connect system (DCS) – A remotely controlled network element based on a microprocessor, which provides the means to connect a time slot in an incoming digital stream to any vacant time slot in an outgoing digital stream. These systems may be used to connect, disconnect, rearrange, and test facilities and private-line services.

Digital Signal Level N, where N may be 0, 1, 2, or 3 – the bit rates at these levels are 64 Kbps, 1.544 Mbps, 6.312 Mbps, and 44.736 Mbps. None of these rates is a direct multiple of a lower level rate, and the signals are asynchronous.

Digital signal cross-connect level N, where N is 1 or 3 – a frame that terminates DS1 or DS3 signals and provides a means for facility cross connection using metallic jumper cables and signal test jacks.

Fiber optics – The branch of optical technology concerned with the transmission of radiant power through fibers made of transparent materials, such as glass, fused silica, or plastic.

Hertz (Hz) – one hertz equals one cycle per second and is the measure of frequency or bandwidth.

Multiplexing – the process of combining a number of individual circuits for transmission over a common transmission path, using either frequency-division or time-division techniques.

Optical carrier level N (OC-N) – the optical equivalent of the STS-N electrical signal.

Optical carrier level N concatenated (OC-Nc) – the optical equivalent of the STS-Nc electrical signal.

Packet data network – a network consisting of processors and communications links, over which data is sent in packet form. A customer occupies a link only during the transmission of the packet.

Pulse code modulation (PCM) – a technique for the transmission of analog information in digital form by sampling the amplitude and encoding the samples using a fixed number of bits.

Permanent virtual circuit (PVC) – a circuit in a packet data network that is set up for a long period.

Reverse direction protection switch (RDPS) – a switch used to reverse the direction of traffic flow around a loop when the loop path is broken.

Synchronous digital hierarchy (SDH) – the European equivalent of SONET.

Synchronous optical network (SONET) – a high-speed synchronous system technology used on fiber-optic cable networks and approved as an international standard in 1988.

Synchronous payload envelope (SPE) – the SPE consists of the payload and the path overhead, and it is carried intact from the point of origin to its destination over a SONET system.

Synchronous transport signal level N (STS-N), where N is an integer – the bit rate at each STS-N level of the SONET multiplexing hierarchy is a multiple of the STS-1 speed of 51.84 Mbps.

Synchronous transport signal level N concatenated (STS-Nc) – the concatenated signal contains only one SPE, which must be kept intact from its point of origin to its destination, and cannot be demultiplexed into multiple lower speed blocks.

Switched virtual circuit (SVC) – a circuit in a packet data network that is set up for the duration of the call.

Synchronization – the technique used to operate digital network elements at the identical clock rate when they are connected by digital transmission facilities.

SYNTRAN – a technique that was developed to provide a synchronous DS3 signal, but was not widely deployed because SONET, which provides a hierarchy of synchronous high-speed digital signals, evolved soon after.

T1 system – a digital transmission system used over metallic pair cables. The T1 system operates at the 1.544-Mbps rate and is commonly used with PCM channel banks to carry 24 VF channels.

Timing jitter – a cumulative impairment in digital systems consisting of a relative timing discrepancy between the elements of digital signals.

Transmultiplexer – a network element that converts blocks of analog channels into blocks of digital channels and vice versa.

Voice frequency (VF) – the voice frequency band is used to carry voice and voiceband data traffic and is usually limited to the 300-Hz to 3,400-Hz frequency range.

Virtual tributary level N – in a SONET system, low-speed traffic is placed in a VT format before being mapped into an SPE.

X.25 – a CCITT standard that defines the packet format for data transfers in a public data network.

List of Acronyms

A

ADM	add/drop multiplexer
ANSI	American National Standards Institute
APS	automatic protection switching
ATM	asynchronous transfer mode

B

B-ISDN	broadband integrated services network
BER	bit-error rate

C

CCITT	Consultative Committee for International Telegraph and Telephone
CO	Central Office
CPE	customer premises equipment/customer provided equipment
CPU	central processing unit
CRC	cyclic redundancy check

D

DCS	digital cross-connect system
DS(N)	digital signal level N, where N may be 0, 1, 2, or 3
DSX-N	digital signal cross-connect level N, where N is 1 or 3

H

Hz   Hertz

O

OC-N   optical carrier level N
OC-Nc   optical carrier level N concatenated

P

PCM   pulse code modulation
PVC   permanent virtual circuit

R

RDPS   reverse direction protection switch

S

SDH   synchronous digital hierarchy
SONET   synchronous optical network
SPE   synchronous payload envelope
STS-N   synchronous transport signal level N, where N is an integer
STS-Nc   synchronous transport signal level N concatenated
SVC   switched virtual circuit

V

VF   voice frequency
VT(N)   virtual tributary level N

Bibliography

ANSI T1.403-1989, Carrier-to-Customer Installation—DS1 Metallic Interface.

ANSI T1X1.3/93-001R2, A Technical Report on Synchronization Management Using Synchronization Status Messages.

AT&T, Technical Reference 54016, Requirements for Interfacing Digital Terminal Equipment to Services Employing the Extended Superframe Format.

AT&T, Technical Reference 62310, Digital Data System Channel Interface Specification.

AT&T PUB 43801, Digital Channel Bank Requirements and Objectives, November 1982.

Bellamy, John, *Digital Telephony*, John Wiley & Sons, Inc., Second Edition, 1991.

Bellcore, Generic Requirements GR-436-Core, Digital Network Synchronization Plan, Issue 1, June 1994.

Bellcore, Generic Requirements GR-2830-Core, Primary Reference Sources: Generic Requirements, Issue 1, March 1994.

Bellcore, Technical Adivsory TA-NWT-001244, Issue 2, Clocks for the Synchronized Network: Common Generic Criteria.

Bellcore, Technical Reference TR-TSY-000012, MML Requirements, January 1985.

Bellcore, Technical Advisory TA-TSY-000055, Basic Testing Functions for Digital Networks and Services, March 1985.

Bellcore, Technical Advisory TA-TSY-000189, Generic Requirements for the Subrate Multiplexer, Issue 1, April 1986.

Bellcore, Technical Advisory TA-TSY-000378, Timing Signal Generator (TSG) Requirements and Objectives, Issue 2, October 1990.

Nicholson, Paul J., "An Overview of the Synchronous Optical Network," *Microwave Journal*, Horizon House Publications, Inc., December 1991.

Thomas, Leonard, "Network Synchronization, DS1 Line Codes, DS1 Frame Formats," SUPERCOMM 90s, May 23, 1989.

Index

The Artech House Telecommunications Library

Vinton G. Cerf, Series Editor

Advanced Technology for Road Transport: IVHS and ATT, Ian Catling, editor

Advances in Computer Communications and Networking, Wesley W. Chu, editor

Advances in Computer Systems Security, Rein Turn, editor

Advances in Telecommunications Networks, William S. Lee and Derrick C. Brown

Analysis and Synthesis of Logic Systems, Daniel Mange

Asynchronous Transfer Mode Networks: Performance Issues, Raif O. Onvural

A Bibliography of Telecommunications and Socio-Economic Development, Heather E. Hudson

Broadband: Business Services, Technologies, and Strategic Impact, David Wright

Broadband Network Analysis and Design, Daniel Minoli

Broadband Telecommunications Technology, Byeong Lee, Minho Kang, and Jonghee Lee

Cellular Radio: Analog and Digital Systems, Asha Mehrotra

Cellular Radio Systems, D. M. Balston and R. C. V. Macario, editors

Client/Server Computing: Architecture, Applications, and Distributed Systems Management, Bruce Elbert and Bobby Martyna

Codes for Error Control and Synchronization, Djimitri Wiggert

Communications Directory, Manus Egan, editor

The Complete Guide to Buying a Telephone System, Paul Daubitz

Computer Telephone Integration, Rob Walters

The Corporate Cabling Guide, Mark W. McElroy

Corporate Networks: The Strategic Use of Telecommunications, Thomas Valovic

Current Advances in LANs, MANs, and ISDN, B. G. Kim, editor

Digital Cellular Radio, George Calhoun

Digital Hardware Testing: Transistor-Level Fault Modeling and Testing, Rochit Rajsuman, editor

Digital Signal Processing, Murat Kunt

Digital Switching Control Architectures, Giuseppe Fantauzzi

Distributed Multimedia Through Broadband Communications Services, Daniel Minoli and Robert Keinath

For further information on these and other Artech House titles, contact:

Artech House
685 Canton Street
Norwood, MA 02062
617-769-9750
Fax: 617-769-6334
Telex: 951-659
email: artech@world.std.com

Artech House
Portland House, Stag Place
London SW1E 5XA England
+44 (0) 171-973-8077
Fax: +44 (0) 171-630-0166
Telex: 951-659
email: bookco@artech.demon.co.uk